ATMOSPHERIC COMPUTATIONS TO ASSESS ACIDIFICATION IN EUROPE

ATMOSPHERIC COMPUTATIONS TO ASSESS ACIDIFICATION IN EUROPE

Summary and Conclusions
of the Warsaw II Meeting

Guest Editors

JOSEPH ALCAMO and JERZY BARTNICKI
IIASA, Laxenburg, Austria

Reprinted from Water, Air, and Soil Pollution
Vol. 40, Nos. 1–2 (1988)

KLUWER ACADEMIC PUBLISHERS
DORDRECHT / BOSTON / LONDON

ISBN-13: 978-94-010-6904-5   e-ISBN-13: 978-94-009-0923-6

DOI: 10.1007/978-94-009-0923-6

Published by Kluwer Academic Publishers,
P.O. Box 17, 3300 AA Dordrecht, The Netherlands.

Kluwer Academic Publishers incorporates
the publishing programmes of
D. Reidel, Martinus Nijhoff, Dr W. Junk and MTP Press.

Sold and distributed in the U.S.A. and Canada
by Kluwer Academic Publishers,
101 Philip Drive, Norwell, MA 02061, U.S.A.

In all other countries, sold and distributed
by Kluwer Academic Publishers Group,
P.O. Box 322, 3300 AH Dordrecht, The Netherlands.

ATMOSPHERIC MODELS AND ACIDIFICATION: SUMMARY AND CONCLUSIONS OF THE WARSAW II MEETING ON ATMOSPHERIC COMPUTATIONS TO ASSESS ACIDIFICATION IN EUROPE

JOSEPH ALCAMO and JERZY BARTNICKI

International Institute for Applied Systems Analysis Schlossplatz 1, A-2361 Laxenburg, Austria

(Received June 1, 1988; revised June 20, 1988)

Abstract. Three topics are discussed in this report: sensitivity/uncertainty analysis of long range transport models, the interface between atmospheric models of different scales, and linkage between atmospheric and ecological models.

In separate analyses of long range transport models, it was found that uncertainty of annual S deposition was mostly affected by uncertainty of wind velocity, mixing height and wet deposition parameterization. Uncertain parameters collectively caused S deposition errors of around 10–25% (coefficient of variation) in the models examined. The effect of interannual meteorological variability on computed annual S deposition was relatively small.

Different methods were presented for combining models of regional and interregional scale. It was found to be more important to include interregional information in regional-scale models for annual computations compared to episodic computations.

A variety of linkage problems were noted between atmospheric and ecological models. The vertical distribution of pollutants and 'forest fittering' of pollutant deposition were found to be important in ecological impact calculations but lacking in the output of most interregional atmospheric models.

1. Purpose of Meeting

One particularly vigorous area in atmospheric research nowadays is the mathematical modeling of meteorological and air quality phenomena. Models resulting from this research serve as tools to not only further scientific understanding, but also to assist in air quality management efforts by helping to identify source-receptor relationships. An important example of this management application is the current role of atmospheric models in assessing the acidification of Europe's environment. A meeting was held in Warsaw 27–29 April, 1987 to address some of the key questions in using atmospheric models to assess acidification in Europe's environment*. The meeting was organized by the International Institute for Applied Systems Analysis (IIASA) in cooperation with the Institute of Meteorology and Water Management (IMGW) in Warsaw. It was a follow-up of a meeting held in Warsaw 4–5 September, 1985 at which preliminary research results covering similar topics were presented. Extended abstracts of the earlier meeting are reported in Alcamo and Bartnicki (1986).

In this summary we review the following subjects: (1) sensitivity/uncertainty analysis of long range transport and deposition models, (2) the interface between atmospheric

---

* A list of participants is presented at the end of this special issue.

Water, Air, and Soil Pollution **40** (1988) 1–7.

models of different scales, and (3) linkage of atmospheric and ecological models. Two other topics were covered in the meeting, 'Update of European Air Pollution Models' and 'Special Report on Air Pollution in Poland', but are not summarized here. This issue of *Water, Air, and Soil Pollution* is devoted to papers from this meeting. In this summary report we review discussions that took place during the meeting, as well as presentations for which written papers are unavailable. We also incorporate some of the results of written papers presented in this issue, though we try to avoid duplication of information.

2. Sensitivity/Uncertainty Analysis of Models

The use of models to evaluate strategies to control transboundary air pollution has brought new importance to the assessment of their uncertainty. Rather than attempt to comprehensively review this issue, as has been done elsewhere (see Fox, 1984; Demerjian, 1985; Carson, 1986), the meeting focused on a few important sub-topics, i.e., different methods of uncertainty analysis and the question of uncertainty due to meteorological variability.

2.1. Methods for Quantifying Model Uncertainty

Eulerian Model Sensitivity Analysis. By their nature, Eulerian air pollution models do not easily yield source-receptor relationships. To address this problem Carmichael* presented a method by which analytically-derived sensitivity equations can be used to establish source-receptor relationships in an Eulerian model. He applied this method to the STEM II SO_2 model of the Eastern U.S. The basic approach is to perturb emissions and then use the sensitivity equations to compute the magnitude of the perturbed SO_2 concentration. These sensitivity coefficients showed a strong diurnal variation at all receptor locations due to changing meteorological situations.

The FAST Method for Sensitivity and Uncertainty Analysis. The topic of sensitivity analysis was also considered in separate papers by Klug and Uliasz. In their studies they used the Fourier Amplitude Sensitivity Test (FAST) to assess model sensitivity and uncertainty, rather than identify source-receptor relationships. The FAST method employs Fourier transforms to limit the sample size required for stochastic simulation. Klug and Erbschauer applied it to a climatological model of SO_2 in Europe and assumed a $\pm 10\%$ of error in model inputs. Through their analysis they found that computed total sulfur deposition was most sensitive to uncertain wind velocity.

Uliasz applied the FAST method to the EMEP I model of SO_2 in Europe. He found for a variety of assumed input uncertainties, that the largest uncertainties in total sulfur deposition were due to (1) the 'mixing height' (which was constant in this model version but time-variable in a later version of the EMEP model) and (2) the parameterization of wet deposition (also revised in the later EMEP model version). When input parameter

---

* A list of presentations to the meeting is available at the end of this special issue.

distributions were assumed to be triangular with $\pm 50\%$ ranges, the computed deposition uncertainty was 17% (coefficient of variation).

Monte Carlo Method of Uncertainty Analysis. Bartnicki used straightforward Monte Carlo Simulation to analyze parameter uncertainty of the EMEP II model. He focused on the relationship between a single source and receptor. Frequency distributions of model inputs were prescribed by a group of atmospheric experts at an earlier meeting. In one model experiment, input parameters were varied every 6 hr; in another, every year. Approximately 1500 annual simulations were run for these experiments. For the 6-hr variations, the computed uncertainty of total annual sulfur deposition was one order of magnitude smaller than for the experiments in which parameter values were varied once a year (2 vs 20%).

In a related presentation, Alcamo studied the parametric uncertainty of the EMEP I model for the case in which several countries (rather than just one) contribute to deposition at a particular receptor. The uncertainty of transfer coefficients between one country and one receptor varied from 11 to 27% (as a coefficient of variation). However, uncertainty was not closely correlated with geographic distance between source and receptor. Instead, there was an inverse proportional relationship between uncertainty and the number of trajectories from a source to a receptor. When the contributions of all countries to a receptor were taken into account, the uncertainty of deposition at Illmitz (Austria) and Rorvik (Sweden) was computed to be 14 and 20%, respectively. This uncertainty estimate increased to 20 and 25% after covariance between transfer coefficients was included.

Comparison of Methods. Uliasz pointed out that the FAST method required a factor of two to four fewer computer runs than Monte Carlo Simulation to perform the same uncertainty calculations of the EMEP I model. Klug also noted the computational efficiency of this method. In addition, both pointed out that partial variances, which are useful in assessing the relative importance of different model inputs, are easily computed with the FAST method. On the other hand, covariances between inputs can be taken into account by Monte Carlo Simulation but not by the FAST method. Moreover, it is difficult to use the FAST method when input frequency distributions are sampled at different time intervals.

Clearly there is no superior approach; the selection of the most appropriate method depends on the problem being addressed.

2.2. METEOROLOGICAL AND CLIMATOLOGICAL VARIABILITY

Model uncertainties can be thought to be (1) reducible, in that model formulation or parameter estimation could be improved, or (2) irreducible, due to inherent variations of nature. One of the most important sources of irreducible uncertainty in these atmospheric models is uncertainty due to interannual meteorological variability and longer-term climate change. This topic was addressed at the meeting in three papers.

Niemann used a simple climatological model of SO_2 in Europe to study the effect of interannual meteorologic variability on computed S deposition. Using meteorological inputs from 1951 to 1985, he found a relatively small interannual variability in wet plus dry deposition. This is supported by the observation reported at the Warsaw I Meeting (Alcamo and Posch, 1986), that interannual meteorological variability between 1979–1982 (in which Grosswetterlagen* frequencies varied substantially) had a relatively small effect on wet plus dry S deposition.

Pitovranov used a variety of approaches to study not only the effects of interannual meteorological variability, but also the effects of potential climate change on S deposition in Europe. In one approach he began by establishing the relationship between hemispheric temperature and regional precipitation measurements. He then used hemispheric temperature increases produced by global circulation models as input to this relationship and computed regional precipitation increases or decreases. He then input these precipitation changes in a simple way to the EMEP I model and examined their effect on sulfur deposition at a few receptors compared to a base case. He found a small effect, though as he pointed out, the EMEP model version he used is not particularly sensitive to the precipitation changes he assumed.

In another approach, Pitovranov compared the frequency of occurrence of Grosswetterlagen between warm and cold climate periods that occurred in this century. He also constructed the frequency distribution of Grosswetterlagen for an artificial year based on parts of selected years. This was based on the correlation between hemispheric temperature and Grosswetterlagen frequency mentioned above. Based on these two Grosswetterlagen approaches, he concluded that global warming would increase blockage of westerly flows, resulting in less transport of pollutants to Eastern Europe, and more frequent northerly flows in spring and less frequent in summer. Following these, some change in long-range transport of pollutants is expected. However, he does not try to estimate the effect of this change on long term S deposition patterns, or their interannual variability.

Den Tonkelaar estimated the frequency of occurrence of different European circulation patterns for a warmer earth by taking three different approaches: (1) examining meteorological data from the five warmest years of this century, (2) using output of a global General Circulation Model, and (3) extrapolating Grosswetterlagen trends from this century to the next. From these three approaches he concluded that the effect of global warming on Europe's circulation patterns would be less than current interannual meteorological differences in these patterns. Hence, deposition differences would also be smaller.

In sum, these papers suggest that the effect of current interannual meteorological variability on S deposition is small (resulting in an average variation in the order of 10 to 20%). This is probably owing to small interannual differences in flow between major emission areas and major deposition areas in Europe. Furthermore, they acknowledge that global warming would affect European general circulation, but they do not agree on how strong this effect will be upon S deposition nor when such an effect would occur. From den Tonkelaar's point of view, the effect of global warming on 'mean' European circulation patterns would be smaller than the current variation of these

---

* Grosswetterlagen are classes of synoptic-scale circulation patterns.

patterns from year to year. Hence, the change in 'mean' annual deposition due to global warming would be smaller than its current interannual deposition.

These papers, and related research, point out the need to quantify the effect of climate change on pollutant deposition patterns. As one possible approach, both Pitovranov and den Tonkelaar recommend that meteorological data from 'artificial meteorological years' (which they have constructed from their Grosswetterlagen analysis) could be used as input to long range transport models.

Furthermore, the following questions remain to be answered: Will global warming affect not only long-term average but also the interannual variation of deposition? Will it be more or less extreme than current interannual variation? What will be more important to the health of forests, lakes, etc... the effect of global warming on pollutant deposition, or the effect of global warming on temperature and precipitation? Another area needing further research is the feedback between pollutants and climate. As an example, changes in Europe's circulation pattern will somewhat alter $SO_4^=$ aerosol patterns which, in turn, may alter local precipitation characteristics.

3. Interface Between Models of Different Scales

To assess acidification and other air pollutant-related problems, atmospheric models of different spatial scales are necessary. For example, in areas distant from major European source areas, interregional models are needed to compute the origin of pollutant deposition. Although the typical spatial resolution of these models (around 100 km) is appropriate for long range transport calculations, it is too coarse for describing the spatial variability of lakes and forests affected by pollutant deposition. Also, near major emission areas, most deposition could come from local sources. Consequently, regional models (with a spatial resolution around 10 to 50 km) are also needed to describe deposition within European countries.

One session of the meeting focused on how to combine information from the inter-regional scale with regional models. Three different methods for connecting these scales were presented to the meeting. In the first case Damrath and Lehmann evaluated the importance of using a coarse-grid interregional model to set boundary conditions for a finer grid regional model of SO_2 in the German Democratic Republic. They showed that it was more important to include interregional information for annual computations of SO_2 than episodic computations. The apparent reason is that episodic concentrations depend more on 'local' sources (i.e., regional sources) than do long-term average concentrations.

Rather than using interregional model results to set boundary conditions, Nordlund superimposed results from the EMEP interregional model on a Finnish regional grid in order to account for annual S deposition due to non-Finnish sources. A regional model was used to compute deposition due to Finnish sources. This approach also provided a convenient way to divide deposition at any Finnish location into Finnish and non-Finnish components. It was noted in the discussion that this superposition may give inaccurate results for calculations with a short (one day, or less) time scale, because of possible nonlinearity between SO_2 emissions and wet deposition at this time scale.

In the third approach to linking interregional and regional time scales, Szepesi used wind sector analysis to establish the 'background' (i.e., non-Hungarian) contributions to both SO_2 and S deposition in Hungary. This background was superimposed on results from a regional model that computed concentration and deposition in Hungary due to Hungarian sources.

In general, these papers, and the discussion following them, emphasized the need to include interregional information in country-scale, regional calculations for *long-term average* SO_2 and S deposition. For calculations with a long time-scale (one year or more), this information can be included by simply super-imposing interregional model results on regional model results or by empirically estimating background values. For episodic computations, interregional information is less important to regional calculations of SO_2 or S deposition. Furthermore, simple superposition may be incorrect because of nonlinearities between emissions and wet deposition at episodic time scales.

The interface issue involves not only using interregional-scale information in regional models, but also the opposite case. Information from country-scale models could be used to estimate aggregated parameters in long range transport models. As an example, a simple parameter is often used in long range transport models to describe the fraction of emissions that is deposited in the grid element of emissions. Though this parameter is usually set spatially constant, in reality it strongly depends on the location of the grid square. Country-scale, regional models could compute the geographic distribution of this parameter within a country and this information can serve as input to long range transport models. This information should reduce the uncertainty of long range transport models because SO_2 concentration is rather sensitive to this parameter (Bartnicki, 1986). Other integrated parameters of long range transport models can be evaluated by country-scale models in a similar way.

4. Linkage Between Atmospheric and Ecological Models

Models that deal with large-scale air pollution problems often include both atmospheric and ecological components. Yet these components are not necessarily compatible with one another. Papers in this session examined some of the issues involved in linking models from different disciplines. Alcamo and Mäkelä studied the problem of using output from an SO_2 atmospheric model as input to a model of SO_2 forest impact. One problem arises because atmospheric models usually do not take into account the vertical gradient of SO_2. Because of this gradient, the SO_2 concentration (annual average) exposed to trees at different elevations was estimated to range from 0.6 to 1.25 times the vertical average at different locations in Europe. Sensitivity analysis of the forest impact model indicated that forest risk calculations were, in fact, very sensitive to this variability. Vertical distributions of pollutants were further discussed by Hakkarinen in an analysis of the extensive SURE data base of aircraft measurements from the U.S. Using these data, the author pointed out that not only SO_2, but also O_3, can have strong vertical gradients. In general, results of atmospheric calculations must be carefully interpreted before being used in forest impact models.

Kämäri elaborated on his earlier paper (Kämäri, 1986) about forest 'filtering' of pollutants. Extensive data were presented from field ecosystems studies that confirmed the importance of accounting for enhanced forest filtering. The ratio of measured forest deposition to bulk deposition in open land ranged from 1.1 to 3.9.

Forest deposition was also the subject of a study by Bredemeier in which he compiled and analyzed data from 10 forest sites in the Federal Republic of Germany. At most sites, acidity of precipitation (i.e., wet deposition rather than total acidic deposition) increases after passing through the tree canopy. This is because the canopy itself collects acidifying gases, particles and droplets. This is the same effect noted by Kämäri. This enhancement of acidity is partly counteracted by buffering of acids in the canopy. However, canopy buffering does not reduce the acid load to soils because trees absorb a quantity of buffering chemicals from soil solution to compensate for the buffering chemicals they use in the canopy. Hence, there is no net buffering of the acid load to soils.

An extensive discussion* followed these papers concerning the kind of output from atmospheric models needed in ecological models. The following data were identified: (1) concentration distributions of SO_2 and NO_x with a finer spatial resolution (< 100 km) than currently available from long range transport models, and a temporal resolution of one month; (2) frequency distributions of sulfur dioxide within each month; (3) historical data about changes in concentration and deposition; (4) case studies about deposition and concentration occurring during selected episodic and short time-scale events.

Acknowledgments

Local organizational details of the 'Warsaw II' meeting were kindly taken care of by Dr J. Pruchnicki.

References

Alcamo, J. and Bartnicki, J. (eds.): 1986, *Atmospheric Computations to Assess Acidification in Europe: Work in Progress*, IIASA Research Report, RR-86-5, Available from International Institute for Applied Systems Analysis, A-2361, Laxenburg, Austria.

Alcamo, J. and Posch, M.: 1986, *Effect of Interannual Meteorological Variability on Computed Sulfur Deposition in Europe*, in J. Alcamo and J. Bartnicki (eds.), IIASA Research Report, RR-86-5, Available from International Institute for Applied Systems Analysis, A-2361, Laxenburg, Austria.

Bartnicki, J.: 1986, *Assessing Atmospheric Model Uncertainty by Using Monte Carlo Simulation*, in J. Alcamo and J. Bartnicki (eds.), *Ibid.*

Carson, D. J.: 1986, *Atmos. Env.* **20**, 1047.

Demerjian, K. L.: 1985, *Bull. Am. Met. Society* **66**, 1533.

Fox, D. G.: 1984, *Bull. Am. Met. Society* **65**, 27.

Kämäri, J.: 1986, *Linkage Between Atmospheric Inputs and Soil and Water Acidification*, in J. Alcamo and J. Bartnicki (eds.), *Ibid.*

* Participants from a parallel meeting on 'Environmental Impact Models to Assess Regional Acidification' took part in this discussion.

RELATIONSHIPS BETWEEN PRIMARY EMISSIONS AND REGIONAL AIR QUALITY AND ACID DEPOSITION IN EULERIAN MODELS DETERMINED BY SENSITIVITY ANALYSIS

SEOG-YEON CHO*, GREGORY R. CARMICHAEL**

Chemical and Materials Engineering, University of Iowa, Iowa City, IA 52242, U.S.A.

and

HERSCHEL RABITZ

Department of Chemistry, Princeton University, Princeton, NJ 06544, U.S.A.

(Received November 17, 1987; revised June 2, 1988)

Abstracts. The relationships between sources, regional air quality and acid deposition are investigated by the use of the sensitivity analysis. A computationally efficient method of calculating sensitivity coefficients is discussed and used to determine source-receptor relations. These techniques are demonstrated using a simplified version of the Sulfate Transport Eulerian Model. The sensitivity analysis is also extended to calculate sensitivities of an objective function by algebraic manipulations of the sensitivity coefficients. The sensitivity coefficients of primary and secondary pollutants with respect to a specific emission are used to study the role of sources in regional air quality. The domain of influence of a source and the maximum value of the response of a receptor region show strong diurnal variations. The source-receptor relations sought by sensitivity analysis show that the ground level sulfate concentration at the receptor region is mainly affected by close-by SO_2 sources during day time and by far-away SO_2 sources during night time. It is also demonstrated that the fate of pollutants emitted in the model region can be found by calculating the sensitivity coefficients of the appropriate objective functions.

1. Introduction

For the past several years, there has been a significant effort to formulate comprehensive Eulerian atmospheric transport/transformation/removal models. The Regional Acid Deposition Model (NCAR, 1986), Acid Deposition Oxidant Model (Venkatram and Karamchandani, 1986) and Sulfate Transport Eulerian Model (hereafter, referred to STEM II model, Carmichael *et al.*, 1986) are examples of the more comprehensive models formulated to date. These models are designed to calculate the distribution of pollutants and acid deposition rates from the source distribution and meteorological fields. However, one of the problems with Eulerian models is that they do not directly provide details of the influence of a specific source at a specific target location (i.e., receptor). Such information on source-receptor relationships is important to the policy maker.

* Present address: Department of Mechanical and Aerospace Engineering, Princeton University, Princeton, NJ 08544, U.S.A.
** Author for all correspondence.

Water, Air, and Soil Pollution **40** (1988) 9–31.
© 1988 *by Kluwer Academic Publishers.*

There have been several attempts to calculate source-receptor relations in Eulerian models. For example, Hsu and Chang (1987) have developed a method for determining source-receptor relations in Eulerian models by including a distinct time-dependent carrier signal on individual sources and decomposing the signal of the pollutant concentrations at specific receptor sites. However, this method requires very accurate numerical techniques so that predicted concentrations can reflect small changes in the emission. Another method has used the idea of 'labeling' sources by adding additional conservation equations for species emitted from different sources (Kleinman, 1988). This technique provides the contributions of individual sources on the secondary pollutant directly produced by these emissions at specific receptor sites. However, this method does not account for effects of variation of emissions on the regional air quality.

In this paper, sensitivity analysis techniques are introduced to evaluate source-receptor relations. The techniques are general and can be used to analyze a variety of related problems. These techniques are demonstrated using the STEM II model with simplified chemistry to calculate emission sensitivities for sources in the Eastern United States during the meteorological conditions on July 4, 1974. The material presented in this paper is restricted to emission sensitivities. However, the same methods can be extended to analyze the physical significance of other model parameters (e.g., the importance of the flux into the model domain, a specific chemical reaction, a specific chemical species, etc.) in determining acid deposition rates.

2. Sensitivity Analysis

The mathematical basis of Eulerian long range transport models is the coupled three dimensional advection-diffusion equation, i.e.,

$$\frac{\partial C_m}{\partial t} + \frac{\partial v_j C_m}{\partial x_j} = \frac{\partial}{\partial x_j} K_{jj} \frac{\partial C_m}{\partial x_j} + R_m + E_m , \tag{1}$$

where C_m is the gas-phase concentration of the mth chemical species, v_j is the wind velocity vector of the jth direction, K_{jj} is the eddy diffusivity tensor, and R_m and E_m denote the chemical reaction term and source term for mth species, respectively, and a summation is implied over the repeated spatial index j. These equations apply to all species. However, short-lived species can be modeled by use of the pseudo-steady state approximation, which results in the algebraic equations,

$$R_m(C_1, \ldots, C_N) = 0 , \tag{2}$$

where N is the number of chemical species. Given the emission and meteorological fields, Equations (1) and (2), with the attended boundary and initial conditions, can be solved to yield the concentration distributions (cf. Carmichael *et al.*, 1986).

The influence of a particular source, $E_l(x'_1, x'_2, x'_3, t)$, on the concentration

$C_m(x_1, x_2, x_3, t)$ can be represented by the quantity

$$\Omega_{m,l}(x_1, x_2, x_3, t; x_1', x_2', x_3', t') = \frac{\delta C_m x_1, x_2, x_3, t)}{\delta E_l(x_1', x_2', x_3', t')} \tag{3}$$

where $\Omega_{m,l}$ is called the 'emission sensitivity density', identified as a functional deriva-
tive. Functional derivatives are used instead of partial derivatives because the emissions
are distributed spatially and temporally within the model domain. This quantity
represents the response of the concentration of the mth species at position (x_1, x_2, x_3)
at time t to an infinitesimal variation of the emissions of the lth species at position
(x_1', x_2', x_3') at time t'. Thus this quantity contains the important 'source-receptor' information. If $\Omega_{m,l}$ is large, then the source of the lth species at (x_1', x_2', x_3') and t' is
important in determining the concentration of the mth species at a receptor located at
(x_1, x_2, x_3) at time t. Also, a positive $\Omega_{m,l}$ means that the concentration of the mth
species increases as the emission of the lth species increases.

The equations describing the densities are derived by taking the first variation of
Equation (1).

$$\frac{\partial \Omega_{m,l}}{\partial t} + \frac{\partial v_j \Omega_{m,l}}{\partial x_j} = \frac{\partial}{\partial x_j} K_{jj} \frac{\partial \Omega_{m,l}}{\partial x_j} + \sum_n \frac{\partial R_m}{\partial C_n} \Omega_{n,l}$$

$$+ \delta_{m,l} \delta(x_1 - x_1') \delta(x_2 - x_2') \delta(x_3 - x_3') \delta(t - t'). \tag{4}$$

Note that even though the species balance equations are nonlinear (e.g., Equation (1)),
the sensitivity density equations are linear. Efficient procedures to calculate these
sensitivity densities have been developed (cf., Cho $et\ al.$, 1987; Reuven $et\ al.$, 1986).

The response of the concentration at a specific receptor to the perturbation of several
species at several locations is represented by

$$\delta C_m(x_1, x_2, x_3, t; v_1' \ldots v_p', t') = \sum_{l=1}^{l=q} \sum_{j=1}^{j=p} \int_{v_j'} \Omega_{m,l} \, \delta E_l(x_1'', x_2'', x_3'', t') \, dv_j'', \tag{5}$$

where p and q are the number of sources and species being perturbed, respectively, and
v_j' is the chosen volume in which the sources are perturbed. If the emissions are
perturbed uniformly (i.e., $\delta E_l(x_1', x_2', x_3', t')$ is constant) then the individual integrals
$\int \Omega_{m,l} \, dv_j'$ represent the importance of source j of the lth species introduced at time t'
on the concentration of the mth species at receptor (x_1, x_2, x_3) at time t, and the relative
importance of the individual sources is given by the relative magnitude of the local
integrals.

In some instances the issue is not how an individual source affects the concentrations
at a given location, but rather the impact of sources in a particular region on a specific
target area. The impact of emissions from the State of Ohio on the ground-level
concentrations in the Eastern United States is a pertinent example. The variation of the
concentrations due to the perturbation of the sources in the region of interest can be

determined by the quantity

$$\delta C_m(x_1, x_2, x_3, t; a_1 < x_1' < b_1, a_2 < x_2' < b_2, a_3 < x_3' < b_3, t')$$

$$= \int_{a_3}^{b_3} \int_{a_2}^{b_2} \int_{a_1}^{b_1} \frac{\delta C_m(x_1, x_2, x_3, t)}{\delta E_l(x_1', x_2', x_3', t')} \, \delta E_l(x_1', x_2', x_3', t') \, dx_1' \, dx_2' \, dx_3' , \qquad (6)$$

where a and b define the boundaries of the source region of interest.

In the case that the normalized variation $\delta E_l(x_1', x_2', x_3', t')/E_l(x_1', x_2', x_3', t')$ is a constant, then the sensitivities are given by

$$S_{m, l}(x_1, x_2, x_3, t; a_1 < x_1' < b_1, a_2 < x_2' < b_2, a_3 < x_3' < b_3, t')$$

$$= \int_{a_3}^{b_3} \int_{a_2}^{b_2} \int_{a_1}^{b_1} \frac{\delta C_m(x_1, x_2, x_3, t)}{\delta \ln E_l(x_1', x_2', x_3', t')} \, dx_1' \, dx_2' \, dx_3' \qquad (7a)$$

$$= D_1(a_1 < x_1' < b_1, a_2 < x_2' < b_2, a_3 < x_3' < b_3, t') \, C_m(x_1, x_2, x_3, t) , \quad (7b)$$

where

$$D_1(a_1 < x_1' < b_1, a_2 < x_2' < b_2, a_3 < x_3' < b_3, t')$$

$$= \int_{a_3}^{b_3} \int_{a_2}^{b_2} \int_{a_1}^{b_1} \frac{\delta}{\delta \ln E_l(x_1', x_2', x_3', t')} \, dx_1' \, dx_2' \, dx_3' . \qquad (8)$$

And $S_{m, l}$ represents the variation of the concentration of species m caused by a fractional and infinitesimal change of the emission of species l. $S_{m, l}$ becomes a normalized sensitivity coefficient if it is divided by $C_m(x_1, x_2, x_3, t)$.

The sensitivity coefficients defined in Equation (7) can be calculated by first applying the operater D defined in Equation (8) to Equation (1). The resulting equation allows the calculation of $S_{m, l}$ directly without first calculating the individual sensitivity densities. The integration over time t' is straightforward. For example, if the perturbation is given from $t = t_i$ to $t = t_f$, then

$$S_{m, l}(x_1, x_2, x_3, t; a_1 < x_1' < b_1, a_2 < x_2' < b_2, a_3 < x_3' < b_3, t_i < t' < t_f)$$

$$= \int_{t_i}^{t_f} \int_{a_3}^{b_3} \int_{a_2}^{b_2} \int_{a_1}^{b_1} \frac{\delta C_m(x_1, x_2, x_3, t)}{\delta \ln E_l(x_1', x_2', x_3', t')} \, dx_1' \, dx_2' \, dx_3' \, dt' \qquad (9a)$$

$$= D_1(a_1 < x_1' < b_1, a_2 < x_2' < b_2, a_3 < x_3' < b_3, t_i < t' < t_f) \, C_m(x_1, x_2, x_3, t) , \quad (9b)$$

where

$$D_l(a_1 < x_1' < b_1, a_2 < x_2' < b_2, a_3 < x_3' < b_3, t_i < t' < t_f)$$

$$= \int_{t_i}^{t_f} \int_{a_3}^{b_3} \int_{a_2}^{b_2} \int_{a_1}^{b_1} \frac{\delta}{\delta \ln E_l(x_1', x_2', x_3', t')} \, dx_1' \, dx_2' \, dx_3' \, dt' . \qquad (10)$$

The analysis can be easily extended to sensitivities of any objective function with respect to emissions. For example, the objective of interest is assumed to be of the following general form

$$\theta = F(\mathbf{C}, \mathbf{E}, x_1, x_2, x_3, t), \tag{11}$$

where F is a functional specified for some particular quantities of interest such as the deposition rate of NO_x or SO_x, or the chemical production rate of HNO_3 or sulfate. The sensitivity coefficients of the objectives are derived as

$$\Omega_{\theta, l}(x_1, x_2, x_3, t; x_1', x_2', x_3', t') = \frac{\delta\theta(\mathbf{C}, \mathbf{E}, x_1, x_2, x_3, t)}{\delta E_l(x_1', x_2', x_3', t')} \tag{12a}$$

$$= \sum_m \frac{\partial F(\mathbf{C}, \mathbf{E}, x_1, x_2, x_3, t)}{\partial C_m(x_1, x_2, x_3, t)} \frac{\delta C_m(x_1, x_2, x_3, t)}{\delta E_l(x_1', x_2', x_3', t')}$$

$$+ \frac{\delta F(\mathbf{C}, \mathbf{E}, x_1, x_2, x_3, t)}{\delta E_l(x_1', x_2', x_3', t')}. \tag{12b}$$

If the normalized variation of the emission is constant over space and time, then

$$S_{\theta, l}(x_1, x_2, x_3, t; a_1 < x_1' < b_1, a_2 < x_2' < b_2, a_3 < x_3' < b_3, t_i < t' < t_f)$$

$$= \sum_m \left(\frac{\partial f(\mathbf{C}, \mathbf{E}, x_1, x_2, x_3, t)}{\partial C_m(x_1, x_2, x_3, t)} \right.$$

$$\left. \times S_{m, l}(x_1, x_2, x_3, t; a_1 < x_1' < b_1, a_2 < x_2' < b_2, a_3 < x_3' < b_3, t_i < t' < t_f) \right)$$

$$+ D_l(a_1 < x_1' < b_1, a_2 < x_2' < b_2, a_3 < x_3' < b_3, t_i < t' < t_f) F(\mathbf{C}, \mathbf{E}, x_1, x_2, x_3, t), \tag{13}$$

where $S_{\theta, l}$ becomes a normalized sensitivity coefficient if it is divided by $\theta(\mathbf{C}, \mathbf{E}, x_1, x_2, x_3, t)$. Once the sensitivity coefficients are calculated, then the sensitivities of objectives can be obtained by algebraic manipulations of sensitivity coefficients as shown in Equations (12) and (13). These techniques are applied to regional scale air quality problems in the following sections.

The approach taken in this section is that the individual sensitivity coefficients are calculated first and then used to obtain the sensitivities of objectives. If the choice of objectives (e.g., the integrated sensitivities over space and time, the variation of concentration at a given location and time, etc.) is pre-determined and the number of objectives is smaller that the number of parameters, then the adjoint method (Hall et al., 1982) may be more economical.

3. Results and Discussions

3.1. DESCRIPTION OF THE MODEL SIMULATION

The sensitivity techniques discussed in the previous section are general and can be applied to any Eulerian model. To illustrate the analysis, a simplified version of the STEM model (Carmichael *et al.*, 1986) for acid deposition is used. The model treats 12 chemical species, of which, $NO_x(NO + NO_2)$, $O_x(NO_2 + NO)$, HNO_3, NH_3, SO_2, sulfate, H_2O_2 are treated as advected species. The remaining species, N_2O_5, NO_3, OH, HO_2, are considered to be short lived species and are modeled by the pseudo-steady state approximation. The mathematical analysis is based on the coupled three-dimensional advection-diffusion equations for the advected species (i.e., Equation (1)). A locally one-dimensional Crank–Nicolson Galerkin finite element technique is used to solve the transport equations and the chemical reaction equations are solved using a pseudo-linearization technique. The details of the solution procedures and the model description can be found in Carmichael *et al.* (1986) and Cho (1986). The model domain and emissions (from the SURE inventory (Hidy *et al.*, 1976)) used in the analysis are presented in Figure 1. The emissions of HNO_3 and H_2O_2 were assumed to be zero and

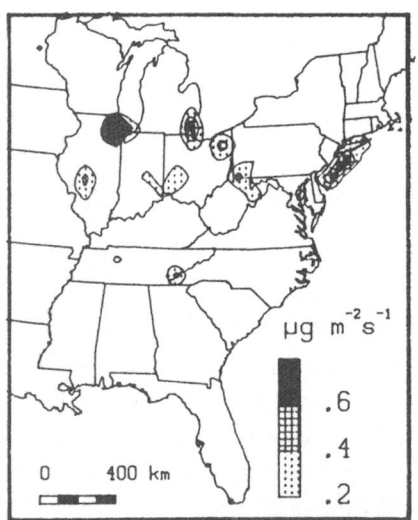 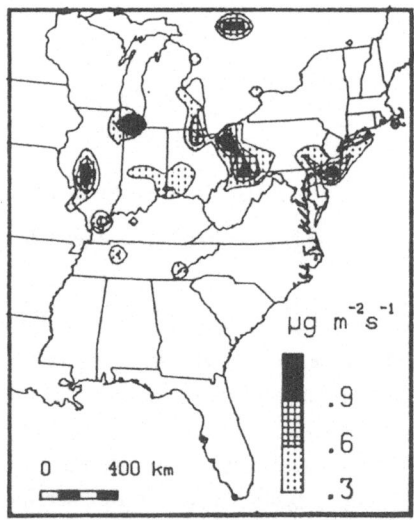

Fig. 1.   The surface NO_x and SO_2 emissions for July 4th, 1974.

the emission of sulfate was assumed to be 0.05 times that for SO_2. All emissions were assumed constant with time. The meteorological fields were from Carmichael and Peters (1984a, b) and varied diurnally. The wind field used is presented in Figure 2. Under the conditions simulated, a high pressure area was located off the coast of North Carolina and a low pressure area was located over Lake Superior. The predicted winds were generally from the west and south-west and increased in magnitude with elevation. The preparation of the inputs is described in detail by Carmichael and Peters (1984a, b).

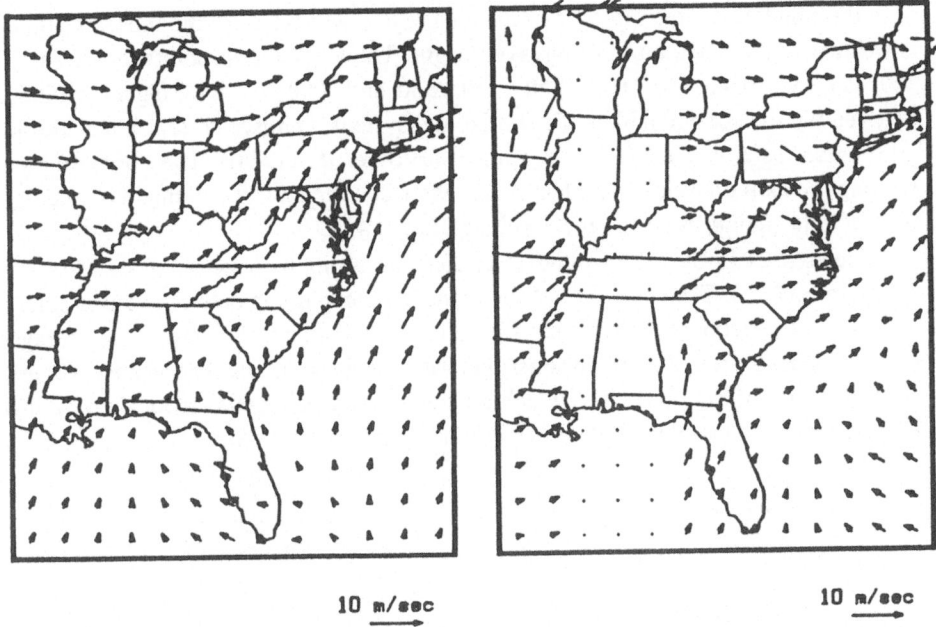

Fig. 2. The surface wind for 00 : 00–12 : 00 (a), and 12 : 00–14 : 00 (b) July 4, 1974.

The initial conditions of SO_2 and sulfate were generated by an inverse r-squared interpolation of observed surface data (c.f., Goodin *et al.* (1979)). The initial concentrations of NO_x, and NH_3 at the surface were assumed to be 5, 55, and 2 ppb, respectively. The initial concentrations of HNO_3 and H_2O_2 were assumed to be zero. The initial vertical profiles of NO_x, O_x, HNO_3, NH_3, H_2O_2, SO_2, and sulfate were determined by the relation

$$C(z) = C(z = 0) e^{-z/H_s}, \qquad (14)$$

where H_s is a function of the properties of the pollutants (e.g., dry deposition velocity and solubility, etc.). The H_s values used in this paper are listed in Table I.

TABLE I

Value of H_s in
Equation (14)

Species	H_s (m)
NO_x	2000
O_x	Infinity
HNO_3	3000
NH_3	2000
H_2O_2	3500
SO_2	2000
Sulfate	3500

In the simulations, the top of the modeling region was set at 3 km, chosen so that it is well above the maximum mixing layer height. The vertical grid points located at 30, 150, 450, 750, 1050, 1350, 1650, 1950, 2250, 2550, and 3000 m provide higher resolution between the surface and 450 m. The horizontal system is chosen so that it includes the grid system used in the SURE experiment (Hidy *et al.*, (1976)). The horizontal grid spacing is 80 km on a side. Simulation runs were conducted from midnight to midnight over a 48 hr period using the meteorology of July 4, 1974.

Calculated 24 hr-average ground level concentrations of selected species are presented in Figure 3. The validation of the model by comparison between measurements and model predictions is not within the scope of this paper. However, the averaged surface concentration profiles are consistent with the observations. The highest averaged concentrations of SO_2 and NO_x, which are the primary pollutants are localized at high emission areas. Also the sulfate and nitric acid, secondary pollutants, are distributed

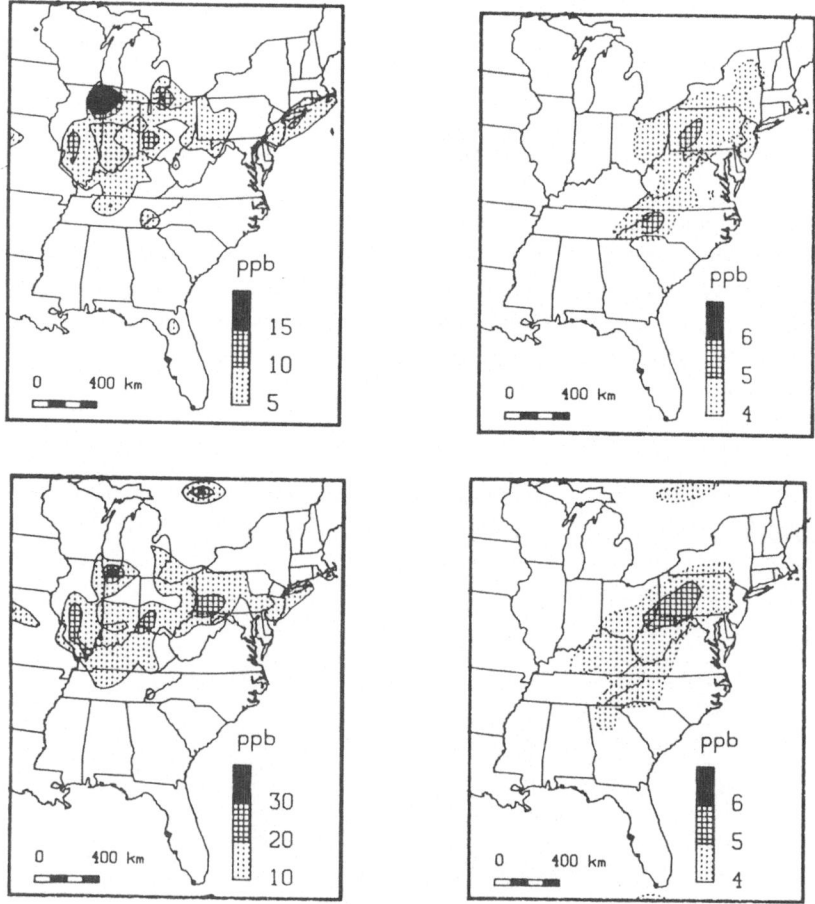

Fig. 3. The calculated 24-hr average concentration of the ground level NO_x, HNO_3, SO_2, and sulfate for July 4, 1974.

more evenly over the model domain. The high concentration regions of sulfate and nitric acid are shifted down wind of the source region of SO_2 and NO_x, and have elongated shapes along the stream lines. This behavior of secondary pollutants suggests that the transport, dry deposition process and chemical kinetics, as well as emission strength, collectively influence the ambient concentration and deposition rates. Due to the fact that the high emission regions of SO_x and NO_x lie in the same locations, except Sudbury, the concentration distributions of SO_2 and sulfate are similar to that of NO_x and HNO_3, respectively.

The total mass balances of NO_x, HNO_3, SO_2, and sulfate in the region are obtained by integrating the appropriate quantities and are presented in Table II. The amount advected is less than 2% of the total inventory for all the advected species. Therefore,

TABLE II

Domain mass balance for the simulation using July 4, 1974 meteorology

	SO_2	Sulfate	NO_x	HNO_3
	——— Unit: 1000 t/model area for 24 hr ———			
The amout present at $t = 0$ hr	117	138	72	0
The amount emitted	110	5	49	0
The amount produced by chemical reaction	− 52	77	− 89	122
The amount advected into the model region	0.4	5	− 4	− 5
The amount deposited	− 16	− 10	− 3	− 11
The amount present at $t = 24$ hr	165	207	28	116

Mass of NO_x is calculated by counting the molecular weight of NO_x as the molecular weight of NO_2.

the uncertainties of boundary conditions have only a slight effect on the predicted concentrations. The amount reacted is larger than the amount deposited from 2 to 10 times depending on the chemical species. Even though the magnitude of the total inventory of SO_2 is comparable to that of sulfate, the amount of sulfate deposited is only half of that of SO_2 due to its smaller dry deposition velocity. The mass balance also indicates that 37% of the sulfate at the end of the 1st day's simulation comes from chemical conversion of SO_2, if the effects of advection and deposition processes are neglected. However, this mass balance analysis does not give any information about how much of the sulfate chemically produced is from SO_2 initially present or from SO_2 emitted. If SO_2 emissions are assumed to be responsible for 48% of the SO_2 in the region, which is the ratio of the amount of SO_2 emitted to the total amount of SO_2 input (i.e., summation of the amount initially present, transported and emitted), then about 18% of the sulfate at the end of the 1st day is from SO_2 emitted. This approximation does not consider the spatial and temporal distribution of the oxidants and the reaction rate constants, nor does it take into account the residence times. For example, the residence time of SO_2 emitted is roughly half of that of SO_2 initially present. A more robust mass balance analysis using sensitivity analysis is discussed in Section 3.5.

3.2. THE EMISSION SENSITIVITY COEFFICIENTS

The influence of a specific source on the regional air quality and the acid deposition can be evaluated by sensitivity analysis. To demonstrate this, the SO_2 source in the vicinity of Gary, Indiana (i.e., source 1 in Figure 4) is selected for investigation. The sensitivity equations (Equation (4)) and the mass balance equations (Equation (1)) were solved to obtain the sensitivities of SO_2 and sulfate with respect to the source of SO_2 in the vicinity of Gary. The perturbation of the emission was assumed uniform relative to emission strength during the entire simulation. In addition, the sensitivity coefficients were normalized by dividing by their own concentrations.

Fig. 4.   Sources and receptors chosen for sensitivity analysis. Sources 1, 2, and 3 are located in the vicinity of Gary, Pittsburgh, and Cincinnati, respectively.

Figure 5 shows the time evolution of the response of the ground level SO_2 to the perturbation of this SO_2 source. Plotted are the percent change in the concentrations (i.e., 100% indicates that a unit magnitude change in the emissions of SO_2 results in a unit magnitude change of predicted concentration). The strong response region shaded in Figure 5 at noontime has an elongated shape in the direction of the wind field (cf. Figure 2 for wind field). A 25% or more response is observed only in the vicinity and down wind of the SO_2 source. This implies that the impact of this SO_2 source is confined to the vicinity of the source area in this case. The distance that SO_2 travels before removal by chemical or physical processes is determined by the chemical reaction rate, wind velocity and dry deposition velocity, and can be estimated by analyzing at the area of the shaded region in Figure 5.

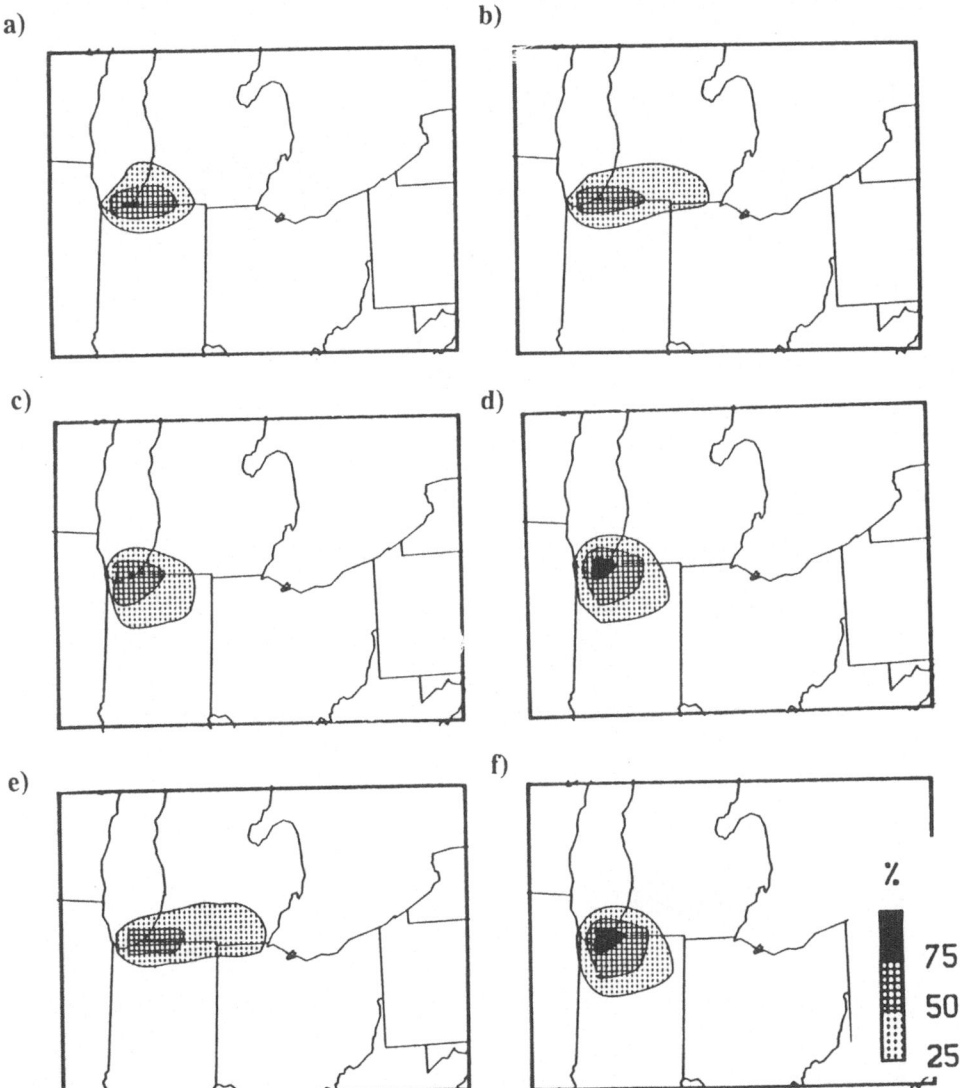

Fig. 5.   The response of the ground level SO$_2$ concentration to the perturbation of the SO$_2$ source in the vicinity of Gary at 06 : 00 (a), 12 : 00 (b), 18 : 00 (c) 24 : 00 July 4 (d), 12 : 00 (e), and 24 : 00 July 5 (f).

The response of the ground level sulfate to perturbations of the SO$_2$ source in the vicinity of Gary is shown in Figure 6. The perturbation of the SO$_2$ source affects sulfate concentration through the photo-chemical reactions. A 3% or more response of sulfate concentration at the surface first appears 6 hr after the simulation starts. This is because the first 6 hr of the simulation was nighttime and the only chemical reaction mechanism for producing sulfate during this period is the heterogeneous pathway, (estimated in the model to be 10^{-6} s^{-1}). As shown in Figure 6b, the source lies at one end of the 3%

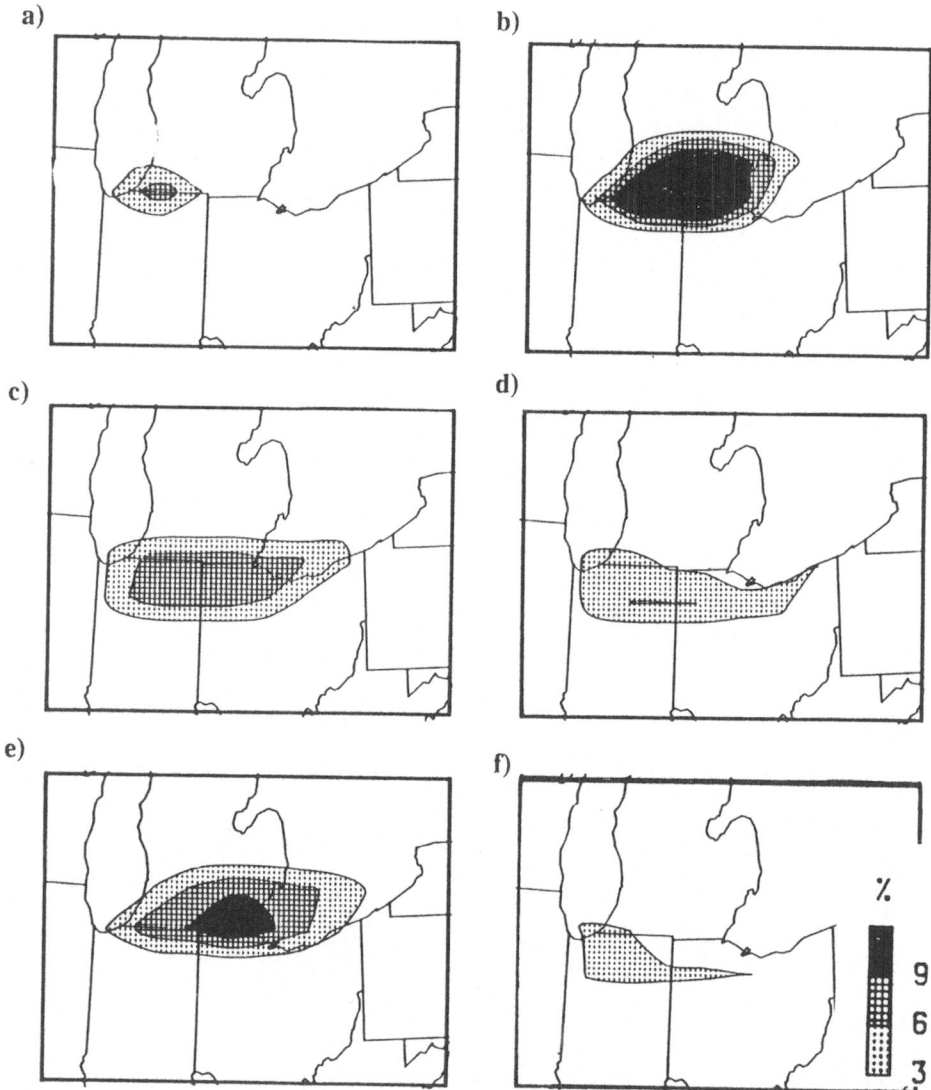

Fig. 6.   The response of the ground level sulfate concentration to the perturbation of the SO₂ source in the vicinity of Gary at 06 : 00 (a), 12 : 00 (b), 18 : 00 (c) 24 : 00 July 4 (d), 12 : 00 (e), and 24 : 00 July 5 (f).

or more response region at noontime but the maximum response region occurs significantly downwind. The sensitivity coefficients for the next 24hr period show similar patterns to those during the first 24 hr as shown in Figure 5 and 6.

The behavior of the sensitivity profiles for SO₂ and sulfate are different. The maximum value of the sensitivities of SO₂ at the surface with respect to perturbation of the SO₂ source increases as time goes from noon to 6 p.m. to midnight. The surface level SO₂ sensitivities increase during the night due to a reduction in the gas phase SO₂ destruction and the limited mixing due to the low mixing layer height during the nighttime. In contrast, the maximum value of the sensitivities of sulfate with respect to the SO_2 source decreases during this time period. This is due to the slower chemical conversion of SO_2 to sulfate during the nighttime.

The diurnal variations of SO_2 and sulfate are clearly shown in Figure 7. The percent variations of the ground level SO_2 and sulfate concentrations along the straight line which is parallel to the abscissa and passes through the perturbation of the SO_2 source

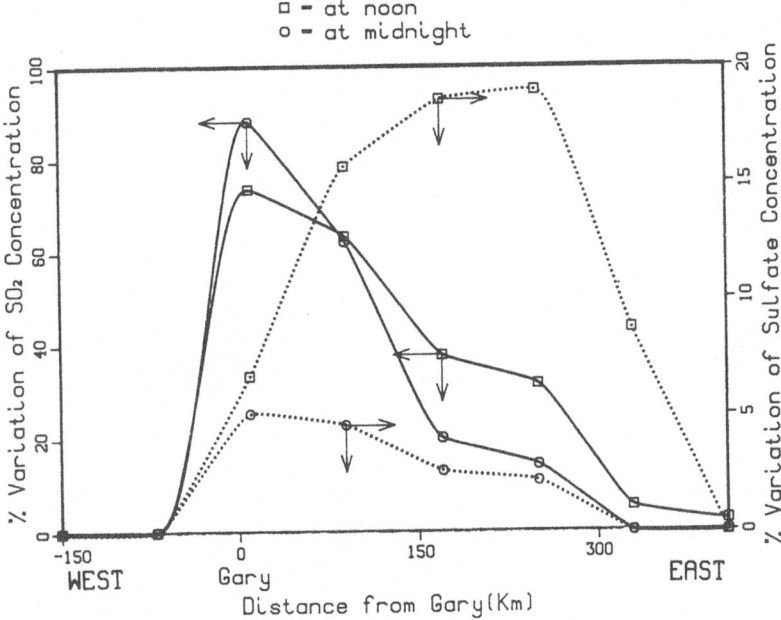

Fig. 7.   The profiles of the percent variations of the ground level SO_2 and sulfate concentrations along the straight line which is parallel to the abscissa and passes through the perturbation of the SO_2 source in the vicinity of Gary.

at Gary, Indiana are presented. The maximum variation of the ground level SO_2 is observed at midnight. The maximum variation of the ground level sulfate is found close to the source region (i.e., Gary, Indiana) at noon and 250 km away from the source region at midnight. This is mainly due to the diurnal variations of the chemical conversion rate of SO_2 to sulfate.

Shown in Figures 8, 9, and 10 are the sensitivities of NO_x, HNO_3, sulfate, and O_3 at the surface with respect to perturbation of the source of NO_x in the vicinity of Gary. The behavior of the sensitivities of NO_x to a NO_x source disturbance is similar to that of the sensitivities of SO_2 to a SO_2 source disturbance (compare Figure 5 with Figure 8). The high response region of NO_x is limited to the vicinity and is downwind of the NO_x source region and the maximum value of the response at nighttime is larger than that during the daytime.

The HNO_3 curves show somewhat different behavior than those for sulfate. For example, the maximum response region of the surface HNO_3 concentration to the

a)

b)

c)

d)

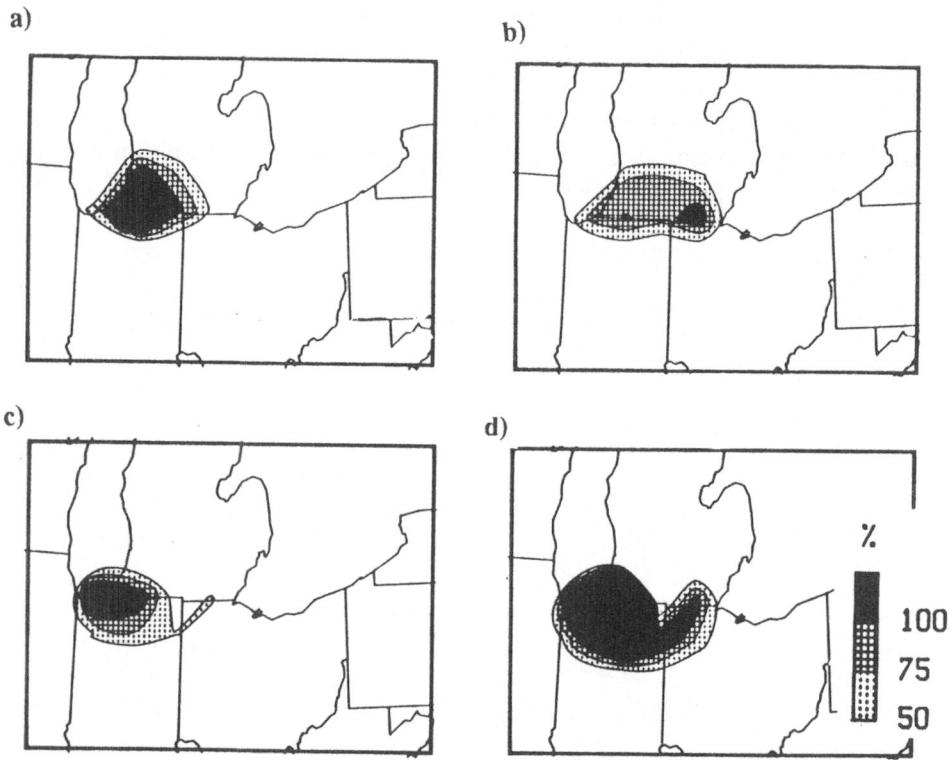

Fig. 8.   The response of the ground level NO_x concentration to the perturbation of the NO_x source in the vicinity of Gary at 60 : 00 (a), 12 : 00 (b), 18 : 00 (c), 24 : 00 July 4 (d).

perturbation of the NO_x source is closer to the source region at nighttime. This is because the nitric acid production rate during the nighttime is faster than that during the daytime for this simulation. The maximum value suggested by Atkinson *et al.* (1982) was used for the reaction rate constant for the $N_2O_5 + H_2O$ reaction, as a result the chemical production rate of HNO_3 is faster by 2 to 10 times at nighttime than at the daytime for this simulation. This is shown in Figure 9.

The responses of the ground level O_3 and sulfate to a perturbation of the NO_x source are shown in Figure 10. Both species show a decrease in concentration due to a positive perturbation of the NO_x source. The negative response of the ground level sulfate to the positive perturbation of the NO_x source is due to the negative response of O_3. O_3 is a percursor of OH which oxidizes SO_2 to sulfate. In contrast, HNO_3, O_3, and NO_x are not sensitive to a perturbation in the SO_2 source.

The influence of simultaneous perturbation of both the NO_x and SO_2 sources upon the ambient concentrations can also be quantified by using Equation (5). The calculated responses of ground level sulfate to perturbation of the NO_x and SO_2 sources in the vicinity of Gary are presented in Figure 11. The response of the ground level sulfate concentration is negative close to the source and becomes positive away from the source.

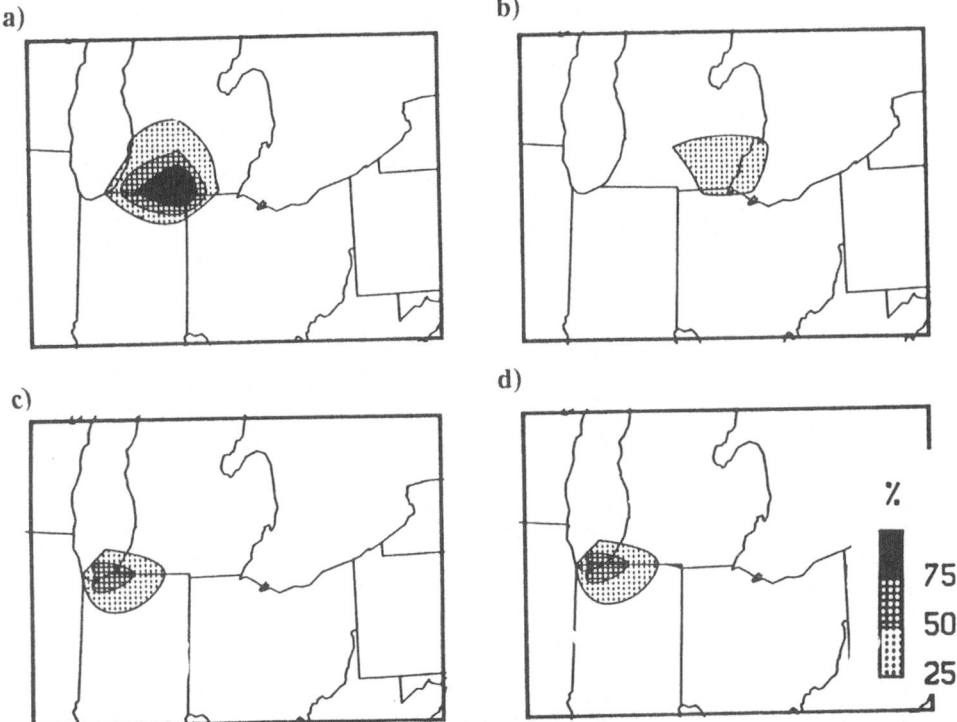

Fig. 9.   The response of the ground level HNO_3 concentration to the perturbation of the NO_x source in the vicinity of Gary at 06 : 00 (a), 12 : 00 (b), 18 : 00 (c), 24 : 00 July 4 (d).

3.3. THE SOURCE-RECEPTOR RELATIONS

The influence of various sources at a receptor location can be evaluated by comparing the sensitivity coefficients at the receptor with respect to the emissions. This is demonstrated for the situation shown in Figure 4. Selected are the SO_2 sources at three different locations in the vicinities of Gary, Pittsburgh, and Cincinnati. Two receptors are chosen to monitor the effect of perturbations to these three different SO_2 sources (see Figure 4 for the location of the chosen receptors). In the analysis that follows, the normalized perturbation ($\delta E/E$) given to each emission was assumed to be uniform.

The sensitivity coefficients of the surface concentration of sulfate with respect to emissions of SO_2 at the three different locations are shown in Figure 12. The diurnal variation of the sensitivity coefficients reflects temporal and spatial changes in the chemical reaction rate, wind velocity, eddy diffusivity, and dry deposition velocity fields. The concentration of sulfate at the surface responds strongly to the close-by SO_2 sources (e.g., the source in the vicinity of Cincinnati for receptor 1 and the source in the vicinity of Pittsburg for receptor 2) during the daytime and the far-away sources (e.g., the source in the vicinity of Gary with respect to receptor 1 and 2) at the nighttime. This is because the distance between the source and the high response region changes as the chemical reaction rates change with times as discussed previously.

a)

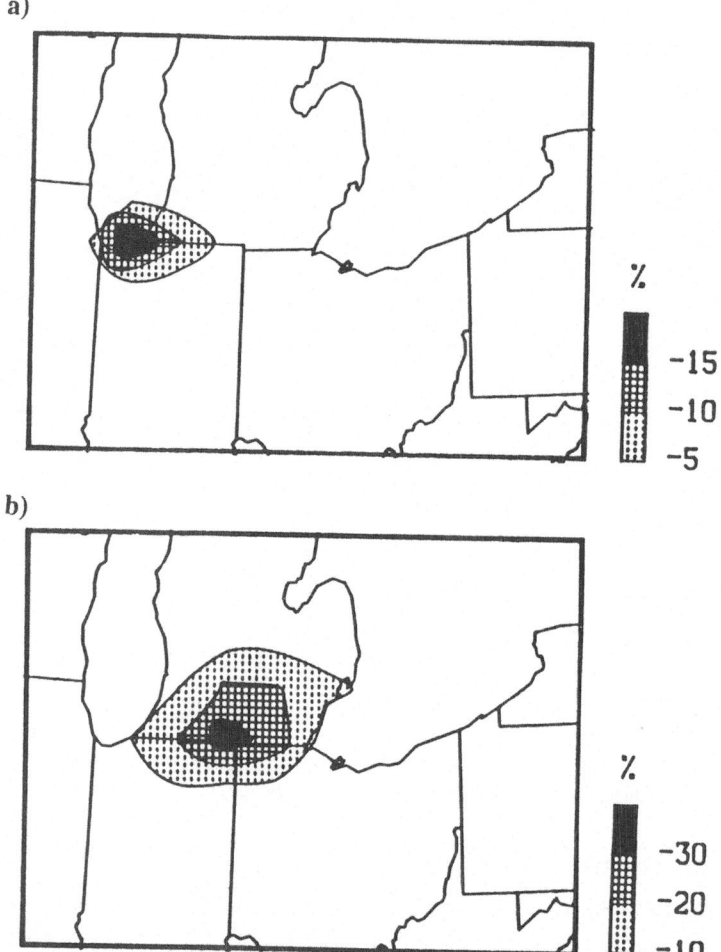

b)

Fig. 10.   The response of the ground level O$_3$ (a) and sulfate (b) concentrations at noon (12 : 00) to the perturbation ot the NO$_x$ source in the vicinity of Gary.

Often times, the receptor and source regions of interest may include more than one grid point (e.g., a specific resource or political region). For example, consider the situation shown in Figure 13 where the source region includes 9 grid points. The source-receptor relationships of these areas can be easily obtained by integrating the emission sensitivity coefficients over the receptor regions. The sensitivity coefficients of the ground level sulfate concentration at receptor A and B with respect to perturbations of the emissions in Ohio (the shaded region in Figure 13) are presented in Figure 14. The diurnal variation of sensitivity coefficients is once again observed. The maximum response of receptor A occurs around 2 PM, and the response of receptor B first appears around 6 AM.

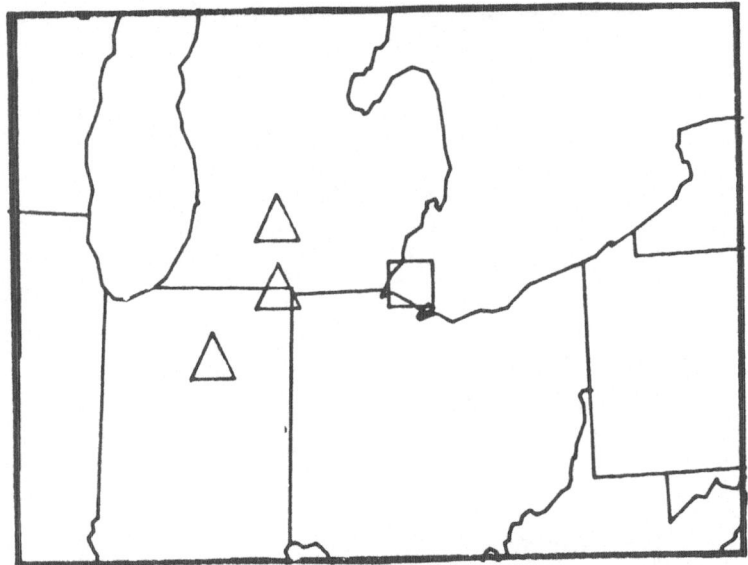

Fig. 11.   The response of the ground level sulfate concentration to a positive perturbation of both SO_2 and NO_x sources in the vicinity of Gary. □ represents positive responses larger than 5% and △ represents negative responses smaller than -5%.

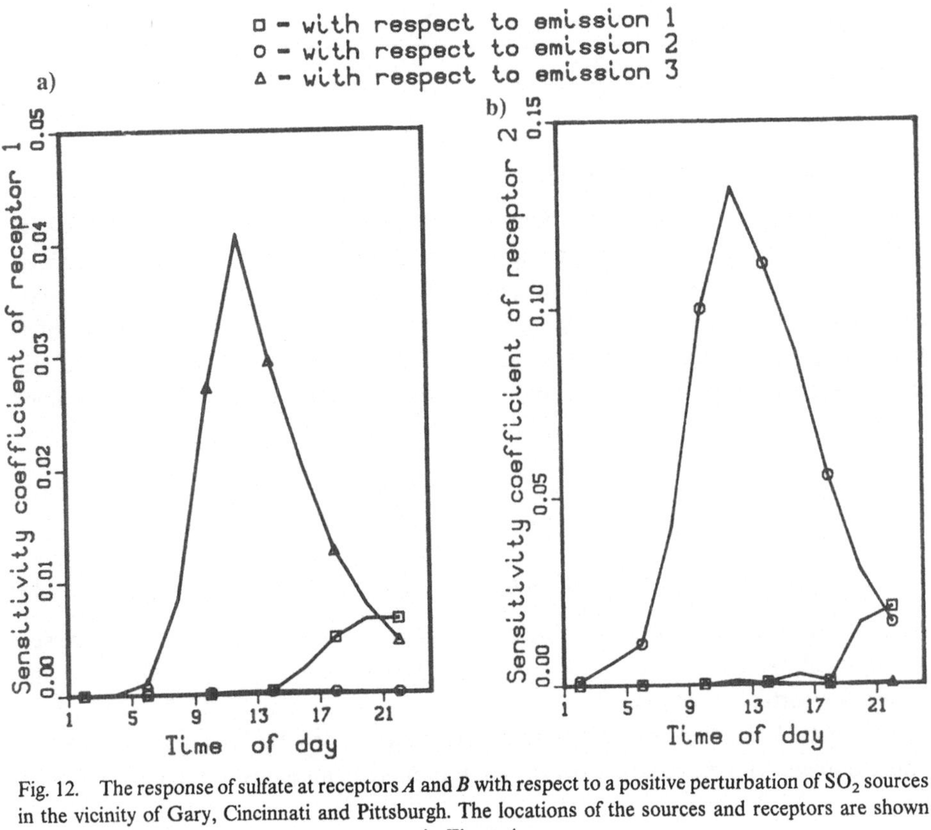

□ - with respect to emission 1
o - with respect to emission 2
△ - with respect to emission 3

Fig. 12.   The response of sulfate at receptors A and B with respect to a positive perturbation of SO_2 sources in the vicinity of Gary, Cincinnati and Pittsburgh. The locations of the sources and receptors are shown in Figure 4.

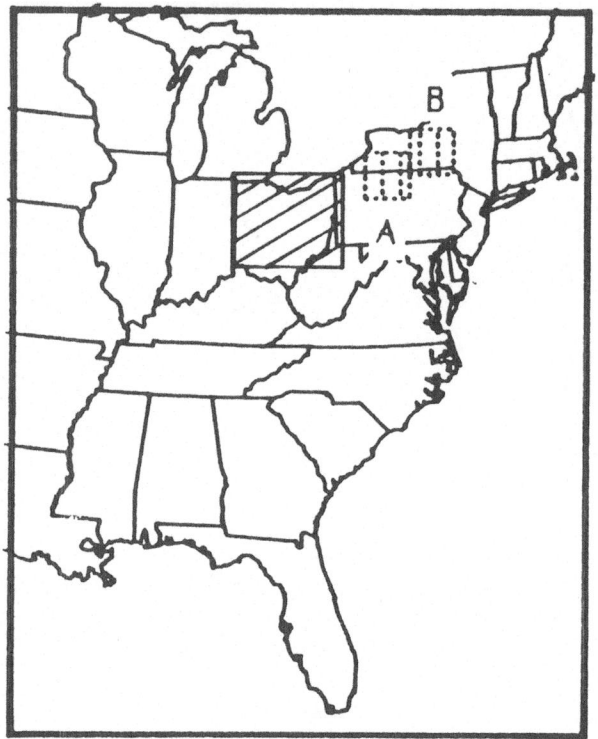

Fig. 13.   The receptor regions selected to monitor the influence of the emissions in Ohio (shown as a shaded region).

3.4. THE RELATIONS BETWEEN PRIMARY EMISSIONS AND AIR QUALITY

The effect of the reduction of emissions and subregions of domain (e.g., one or several states) on the regional air quality can be quantified by the sensitivity coefficients defined in Equations (7) and (9). In this section, perturbations are given to emissions in the State of Ohio and in the entire model region, respectively. The spatially distributed sensitivity coefficients are 'lumped' into the average value by integrating the sensitivity coefficients in the horizontal directions. In this study, the following 'lumped' sensitivities are defined

$$\gamma = \frac{1}{A} \int_A [S_{m,l}(x_1, x_2, x_3, t; a_1 < x'_1 < b'_1, a_2 < x'_2 < b_2,$$

$$a_3 < x'_3 < b_3, t_i < t' < t_f)/C_m(x_1, x_2, x_3, t)] \, dx_1 \, dx_2, \tag{15}$$

where $S_{m,l}(x_1, x_2, x_3, t; a_1 < x'_1 < b_1, a_2 < x'_2 < b_2, a_3 < x'_3 < b_3, t_i < t' < t_f)$ are sensitivity coefficients defined in Equation (9), A is the area of the model region, and x_1 and x_2 are the surface directions.

The lumped sensitivities of the ground level sulfate concentration integrated over the entire domain with respect to a perturbation of all the emissions in the model domain

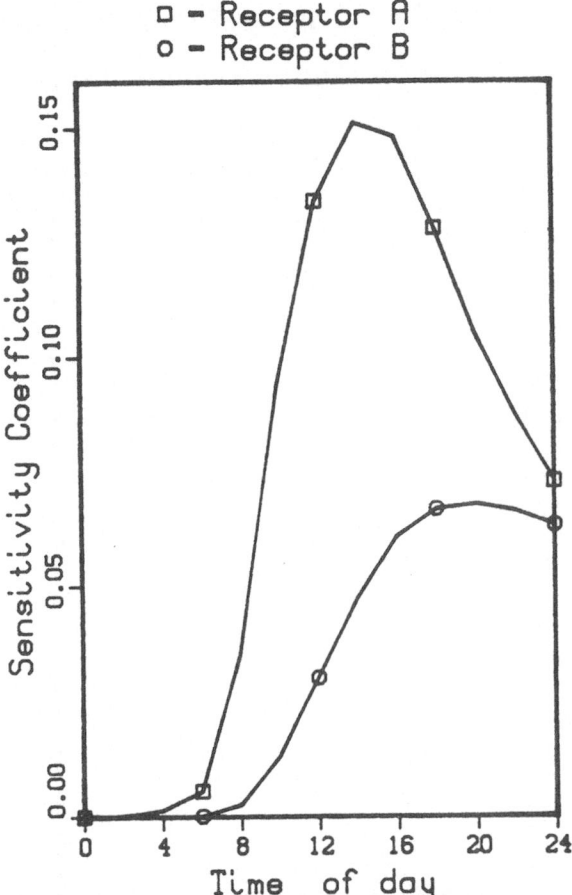

Fig. 14.   Sensitivity coefficients of the ground level sulfate with respect to perturbation of emissions in Ohio at the two receptor regions shown in Figure 13.

and to a perturbation of the emissions in Ohio vs time are presented in Figure 15. The largest response of the sulfate concentration at the surface is observed at noontime for both cases. The sensitivity value for the case that all the sources in the model domain are perturbed is relatively small; the maximum response is less than 5% of the perturbation of the sources. This response is smaller by three times or so than the rough estimates of the contribution of the SO_2 emission to the amount of sulfate at the end of the first day of the simulation based on mass balance (cf., Section 3.1). The sensitivity coefficients of sulfate when only the sources in Ohio are perturbed are ≅ 6 or 7 times smaller than that when all the sources are perturbed. The ratio of the maximum value of the sensitivity coefficients of sulfate with respect to the Ohio emissions to that of the total emissions (i.e., in the Eastern United states) is the same as the ratio of SO_2 emissions of Ohio to the total SO_2 emissions.

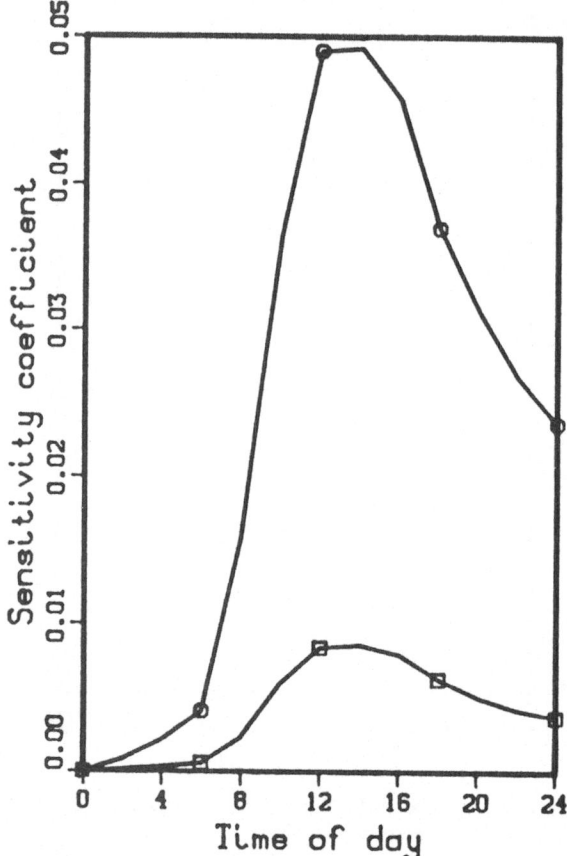

□ ▬ **Perturbation of the emissions in Ohio only**

○ ▬ **Perturbation of all emissions in the domain**

Fig. 15. The lumped sensitivity coefficients of the total domain averaged ground level sulfate concentrations with respect to perturbations of SO_2 emissions in Ohio and perturbations to all the SO_2 emissions in the model domain.

3.5. SENSITIVITY ANALYSIS OF MASS BALANCE

Contributions of sources to the air quality of the model domain or surroundings can be further assessed by calculating the sensitivity coefficients of the terms in the domain mass balance. The total mass of the nth species after time t is described as

$$M_n(t) = M_n(t_0) - FO_n(t) + FI_n(t) - VG_n(t) + R_n(t) + EM_n(t),   (16)$$

where $M_n(t)$ and $M_n(t_0)$ are the total mass of the nth species in the region at time t and $t = 0$ respectively. $FO_n(t)$ and $FI_n(t)$ denote the total amount of the nth species advected out from the boundary and into the boundary, respectively, $VG_n(t)$ and $EM_n(t)$ represent the total amount of nth species deposited and emitted respectively, from time 0 to t, and

$R_n(t)$ is the total amount of the nth species produced by chemical reaction from time 0 to t.

The sensitivities of the individual terms in Equation (16) with respect to the emissions can be obtained by algebraic manipulation of the sensitivity coefficients of the chemical species with respect to emissions by using Equation (13). Assuming that the chemical species advected into the boundary do not reflect the change of the concentration inside of boundary, then

$$D_1^0(M_n(t) + FO_n(t) + VG_n(t) - R_n(t)) = EM_n(t) \delta_{n,l}, \tag{17}$$

where

$$D_1^0 = \int_0^t \int_{v'} \frac{\delta}{\delta \ln E_l(x_1', x_2', x_3', t')} \, dv' \, dt', \tag{18}$$

where v' is the model domain. If $l = n$, sensitivities of each term in Equation (16) with respect to emission, $E_{l=n}$, can be used to interprete the fate of the nth species emitted from the source. For example, $(D_n^0 M_n(t))/EM_n(t)$ represents how much of the nth species emitted from time 0 to t remains in the model domain at time t. Similarly, how much of the nth species removed by chemical reaction, dry deposition or advection can be estimated by $(-D_n^0 R_n(t))/EM_n(t)$, $(-D_n^0 VG_n(t))/EM_n(t)$ and $(-D_n^0 FO_n(t))/EM_n(t)$, respectively. The contribution of emissions of the lth species on the total inventory at t and the amount reacted, deposited and advected for the nth species can also be evaluated by calculating $(D_l^0 M_n(t))/M_n(t)$, $D_l^0 R_n(t)$, $(D_l^0 VG_n(t))/VG_n(t)$ and $(D_l^0 FO_n(t))/FO_n(t)$, respectively. For example, for the conditions simulated $(D_n^0 M_n(t))/EM_n(t)$ and $(-D_n^0 R_n(t))/EM_n(t)$ are 0.84 and 0.14, respectively, and $(D_n^0 VG_n(t))/EM_n(t)$ and $(D_n^0 FO_n(t))/EM_n(t)$ are negligible when $n = SO_2$. This indicates that 84% of the SO_2 emitted remains in the domain in the gas phase and 14% of it is converted to sulfate at the end of the first day of simulation. The calculation of $(D_l^0 M_n(t))/M_n(t)$, $(-D_l^0 R_n(t))/R_n(t)$ for $n, l = SO_2$ shows that 56% of the SO_2 in the model region after 24hr and 30% of the amount of sulfate produced chemically by the SO_2 are from the SO_2 emitted. This analysis also indicates that 11% of the amount of the sulfate in the model region after 24 hr is from the SO_2 emitted. These estimations of the contribution of the SO_2 emission on the SO_2 and sulfate inventory and the amount of the SO_2 and sulfate chemically produced should be more accurate than those obtained from the qualitative analysis of the mass balance in Section 3.1.

4. Summary

Techniques for use in investigating the relationships between sources and regional air quality were introduced and discussed. The techniques allow the efficient calculation of the sensitivity coefficients with respect to spatially distributed emissions. These methods are general and can be applied to perform the sensitivity analysis with respect to any other parameters.

A simplified version of STEM II was used to demonstrate the use of sensitivity analysis for Eulerain Models. The sensitivity coefficients of pollutants with respect to emissions in a specific region indicate the variation of the pollutants to perturbations of the sources. Thus the region and magnitude of influence of a source can be obtained by calculating the sensitivity coefficients of pollutants with respect to the source. The response of SO_2 at the surface to the disturbance of the SO_2 source at Gary, Indiana was calculated and found to have a maximum value at nighttime while that of sulfate to the disturbance of this SO_2 source had a maximum during the daytime. The perturbation of the NO_x source at this location was shown to affect not only NO_x and HNO_3 but also O_3 and sulfate. The simultaneous reduction of NO_x and SO_2 in Gary resulted in the decrease of the ground level sulfate concentration in close-by receptors and the increase of the ground level sulfate concentration at far-away receptors.

Source-receptor relationships were also obtained by recording the emission sensitivity coefficients at different receptor points or areas. The ground level sulfate concentrations at the receptor regions were sensitive to close-by sources during daytime and by far-away sources during the nighttime. The importance of the emissions from the State of Ohio on the air quality of the Eastern United States was also investigated by 'lumping' the sensitivity coefficients. Also the fate of chemical species emitted in the model domain was determined by calculating the sensitivity coefficients of the total inventory at time t and the amount reacted and advected of chemical species with respect to emissions.

The sensitivity analysis tools introduced and demonstrated in this study provide extremely useful tools for quantifying specific issues regarding source-receptor relations and the role of physical and chemical processes in Eulerian models. We are currently appying these sensitivity analysis techniques to a more complete version of STEM II which describes not only gas phase processes but also liquid phase processes.

Acknowledgments

Computer support was provided by the Graduate College, University of Iowa. Author Herschel Rabitz would like to acknowledge the Department of Energy, and authors Cho and Carmichael would like to acknowledge the Electric Power Research Institute and PRECP program (through Battelle Pacific Northwest Laboratory) for partial support of this research.

References

Atkinson, R., Lloyd, A., and Winges, L.: 1982, *Atmos. Env.* **16**, 1341.
Carmichael, G. R. and Peters, L. K.: 1984a, *Atmos. Env.* **18**, 937.
Carmichael, G. R. and Peters, L. K.: 1984b, *Atmos. Env.* **18**, 953.
Carmichael, G. R., Peters, L. K., and Kitatda, Toshihiro: 1986, *Atmos. Env.* **20**, 173.
Cho, S. Y.: 1986, 'Mathematical Modeling and Sensitivity Analysis of Acid Deposition', Ph. D. Thesis, University of Iowa.

Cho, S. Y., Carmichael, G. R., and Rabitz, H.: 1987, *Atmos Env.* **21**, 2589.

Goodin, W. R., McRae, G. J., and Seinfeld, J. H.: 1979, *J. Applied Meteorology* **18**, 761.

Hall, M. C. G., Cacuci, D. G., and Schlesinger, M. E.: 1982, *J. Atmos Sci.* **39**, 2038.

Hidy, G. M., Tong, E. Y., Mueller, P. K., Rao, S., Thomson, F., Berlandi, F., Muldoon, D., McNaughton, D., and Majahad, A.: 1976, *Design of the Sulfate Regional Experiment*, PB, pp. 251–701.

Hsu, H. M. and Chang, J. S.: 1987, *J. of Atm. Chem.* **5**, 103.

Kleinman, L. I.: 1988, *Atmos. Env.*, (in press).

NCAR: 1986, *Preliminary Evaluation Studies with the Regional Acid Deposition Model*, National Center for Atmospheric Research, ADMP-68-4.

Reuven, Y., Smooke, M. D., and Rabitz, H.: 1986, *J. Comp. Phys.* **64**, 27.

Venkatram, A. and Karamchandani, P.: 1986, *Environ. Sci. Techno.* **20**, 1084.

APPLICATION OF THE FAST METHOD TO ANALYZE THE SENSITIVITY-UNCERTAINTY OF A LAGRANGIAN MODEL OF SULPHUR TRANSPORT IN EUROPE

MAREK ULIASZ

Institute of Environmental Engineering, Technical University of Warsaw, Nowowiejska 20, 00-653 Warsaw, Poland

(Received November 17, 1987; revised May 8, 1988)

Abstract. A Fourier Amplitude Sensitivity Test (FAST) is applied to study uncertainty of the EMEP-W atmospheric model of long range transport of S in Europe. The FAST method requires frequency distribution of model parameters as input data and provides the following results: (i) mean value of the model output, (ii) variance of the model output which characterizes the parameter uncertainty of the model, and (iii) partial variances of the model output which are the measures of the model sensitivity to uncertainties in individual input parameters. A mathematical formulation of the FAST method and approximations used in its computer implementation is presented. The application requires an extension of the original method to evaluate the model output frequency distribution for different forms of prescribed frequency distributions for parameters. The computer program allows one to apply the FAST method for an arbitrary mathematical model with different options defined by the user. Some examples of uncertainty analysis of the EMEP model using real meteorological and emission data are described and compared with results obtained by the Monte-Carlo method.

1. Introduction

A comprehensive framework for error analysis is proposed for the EMEP-W model of long range atmospheric transport of S in Europe by Alcamo and Bartnicki (1985, 1987). This framework includes an assessment of parameter uncertainty of the model with the aid of the Monte-Carlo method which allows one to evaluate frequency distributions of model outputs for assumed frequency distributions of input parameters. However, it does not help one to understand what drives the model uncertainty since contribution of uncertainties of individual input parameters cannot be simultaneously identified. Furthermore, the Monte-Carlo method is time-consuming on the computer because a large number of model solutions is required to obtain a statistically significant sample.

The present work applied the Fourier Amplitude Sensitivity Test (FAST) as an alternative to Monte-Carlo approach for investigating parameter uncertainty of the long range transport model. The FAST method was originally developed by Cukier *et al.* (1978) as a technique for sensitivity analysis of mathematical models and it provides important information on model sensitivity to uncertainties in individual parameters that is not available with the Monte-Carlo approach. The FAST method requires frequency distributions of model parameters as input data and gives the following results:

(i) mean values of the model outputs;

(ii) variances of the model outputs which characterize model uncertainty due to uncertainties in input parameters;

Water, Air, and Soil Pollution **40** (1988) 33–49.
© 1988 by Kluwer Academic Publishers.

(iii) partial variances of model outputs which are the measures of model sensitivity to uncertainties in individual parameters.

FAST has been successfully applied for different models including a chemical laser model, a chemical reaction model and an economic model (Cukier *et al.*, 1978), enzyme mechanisms (Pierce *et al.*, 1981), photochemical air polution models (Falls *et al.*, 1979; Tilden and Seinfeld, 1982) and atmospheric boundary-layer models by the author. A review of FAST and another sensitivity analysis methods can be found in Koda *et al.* (1979) and Uliasz (1985). The second work discusses applicability of various approaches (indirect, direct, variational methods, and FAST) to study sensitivity of complicated meteorological models with distributed parameters. A special emphasis is given to interpretation of results available from the linear methods and the nonlinear (or global) sensitivity methods like FAST or Monte-Carlo approach.

Mathematical formulation of the FAST method and approximations used in its computer implementation are briefly described below following mainly Cukier *et al.* (1978) including some extensions required by the specific application. Special attention is devoted to evaluation of output frequency distribution for different forms of prescribed frequency distributions for parameters. The computer program allows one to use the FAST method in the same manner as the Monte-Carlo analysis. Some examples of uncertainty analysis of the EMEP model using real meteorological and emission data are described and comparison with results from the Monte-Carlo analysis is discussed.

2. Mathematical Formulation

Let us consider the general mathematical model

$$\mathbf{Y} = \mathbf{F(X)}, \tag{1}$$

where $\mathbf{X} = \{x_i, i = 1, \ldots, m\}$ is a vector of model input parameters, $\mathbf{Y} = \{y_j, j = 1, \ldots, n\}$ is a vector of model outputs and $\mathbf{F}$ is an operator acting on $\mathbf{X}$. The ensemble mean for the model output y_j is given by

$$\langle y_j \rangle = \int \ldots \int y_j(x_1, \ldots, x_m) \mathbf{p}(x_1, \ldots, x_m) \, dx_1 \ldots dx_m, \tag{2}$$

where $\mathbf{p}$ is the m-dimensional probability density function for $\mathbf{X}$. The central idea of the FAST method is to convert the above integral over the m-dimensional space of input parameters into a one-dimensional integral over a certain search variable s:

$$\langle y_j \rangle = \lim_{T \to \infty} \frac{1}{2T} \int_{-T}^{T} y[x_1(s), \ldots, x_m(s)] \, ds. \tag{3}$$

It is done by assigning a frequency ω_i to each input parameter x_i and using the transformation

$$x_i = G_i[\sin(\omega_i s)], \quad i = 1, \ldots, m \tag{4}$$

under the assumption that the following conditions are satisfied:

(i) the input parameters are assumed to be uncorrelated, i.e., their probability density functions are independent

$$\mathbf{p}(x_1, \ldots, x_m) = \prod_{i=1}^{m} p_i(x_i) \, ; \tag{5}$$

(ii) the frequency set $\{\omega_i\}$ is incommensurate, i.e.,

$$\sum_{i=1}^{m} \gamma_i \omega_i = 0 \, , \tag{6}$$

where the values γ_i are arbitrary integers;

(iii) the functions G_i are chosen so that the arc length, ds, is proportional to $\mathbf{p}(x_1, \ldots, x_m) \, dx_i$ for all i.

The transformation functions (Equation (4)) describe a search curve that samples the parameter space in a manner consistent with the statistics expressed by $\mathbf{p}(x_1, \ldots, x_m)$. Specifying one value of search variable s specifies all values of the parameters $x_1, x_2, \ldots, x_m$. As s varies from $-\infty$ to $+\infty$ all possible x_i values are obtained via Equation (4). If it were possible to use an incommensurate frequency set the search curve would be infinitely long and would pass arbitrarily close to every point in the parameter space. However, in practice, an integer rather than an incommensurate frequency set must be used. This introduces two types of error:

(i) the search curve is no longer space-filling, i.e., the search curve with integer frequencies is a closed curve that does not pass arbitrarily close to any point in the parameter space (but the use of higher frequencies results in the longer search curve);

(ii) the fundamental harmonics used to described the set $\{x_i\}$ will have harmonics that interfere with one another.

These errors resulting in differences between the integrals (2) and (3) can be made arbitrarily small by a proper choice of the integer frequency set.

The use of integer frequencies in the transformation (4) implies that the input parameters x_i and in turn the model outputs y_j are periodic in s on the interval $[-\pi, \pi]$, i.e., $x_i(s) = x_i(s + 2\pi)$ and $y_j(s) = y_j(s + 2\pi)$. The model outputs can thus be Fourier analyzed to obtain their Fourier coefficients

$$A_k^{(j)} = \frac{1}{2\pi} \int_{-\pi}^{\pi} y_j[x_1(s), \ldots, x_m(s)] \cos(ks) \, ds, \quad k = 0, 1, 2, \ldots, \tag{7}$$

$$B_k^{(j)} = \frac{1}{2\pi} \int_{-\pi}^{\pi} y_j[x_1(s), \ldots, x_m(s)] \sin(ks) \, ds, \quad k = 1, 2, 3, \ldots. \tag{8}$$

The mean value $\langle y_j \rangle$ and the variance σ_j^2 of the model output y_j can be written in terms of Fourier coefficients according to Parseval's theorem:

$$\langle y_j \rangle = \sqrt{2} \, [A_0^{(j)}], \tag{9}$$

$$\sigma_j^2 = \langle y_j^2 \rangle = 2 \sum_{\substack{k=-\infty \\ k \neq 0}}^{\infty} [(A_k^{(j)})^2 + (B_k^{(j)})^2] \,. \tag{10}$$

The total variance does not identify the contributions of the individual parameters to σ_j^2. Therefore, the *partial variances* are introduced by selecting from all the Fourier coefficients those corresponding to the fundamental frequency ω_i and its harmonics $p\omega_i (k = p\omega_i, p = 1, 2, \ldots)$. These coefficients express the contribution of the ith input parameter variation into the total variance σ_j^2 of the model output:

$$\sigma_{j/i}^2 = 2 \sum_{p=1}^{\infty} [(A_{p\omega_i}^{(j)})^2 + (B_{p\omega_i}^{(j)})^2] \,. \tag{11}$$

The normalized sensitivity measure, the partial variance $S_{j/i}$, is then defined by the ratio of the above variance due to uncertainty of the ith parameter to the total variance:

$$S_{j/i} = \sigma_{j/i}^2 / \sigma_j^2 \,. \tag{12}$$

It is apparent that the sum $\sum_{i=1}^{m} S_{j/i}$ of the partial variances will not equal unity because $\sigma_{j/i}^2$ in the numerator of Equation (12) involves only the sum of the squares of the Fourier coefficients of the fundamental and all harmonics of the ith frequency ω_i. The total variance σ_j^2 can be written as the following sum:

$$\sigma_j^2 = \sum_{i=1}^{m} \sigma_{j/i}^2 + \sum_{i=2}^{m} \sum_{l=1}^{i-1} \sigma_{j/il}^2 \sum_{i=3}^{m} \sum_{l=2}^{i-1} \sum_{k=1}^{l-1} \sigma_{j/ilk}^2 + \ldots \,, \tag{13}$$

where the second and further terms contain increasingly more detailed information about the coupling of sensitivity among larger and larger groups of parameter uncertainties. For example, a part of total variance of the jth model output related to the couple effect of uncertainties in ith and lth parameters can be expressed as the following *coupled partial variance*:

$$S_{j/il} = \sigma_{j/il}^2 / \sigma_j^2 = \frac{2}{\sigma_j^2} \sum_{p=1}^{\infty} \sum_{q=1}^{\infty} [A_{p\omega_i + q\omega_l}^{(j)})^2 + (B_{p\omega_i + q\omega_l}^{(j)})^2] \,. \tag{14}$$

When coupled partial variances are large it is difficult to separate the effect of one input parameter from that of the other.

3. Computer Implementation

Application of the FAST method requires the numerical evaluation of the Fourier coefficients (Equations (7) and (8)). This in turn requires the model output y_j be evaluated as the search variable s ranges over $[-\pi, \pi]$. Restricting the frequency set to odd integers reduces the range of s to the interval $[-\pi/2, \pi/2]$ since it is possible to use the symmetry relations for the output functions. The model is solved for r sampling points uniformly spaced along the search curve throughout the range $[-\pi/2, \pi/2]$:

$$s = \frac{\pi}{2} \frac{2l - r - 1}{r} \,, \quad l = 1, 2, \ldots, r \,. \tag{15}$$

Each value of the search variable s is related to the input parameters x_i by the transformation (4).

The Fourier coefficients may be now written in the finite difference form using a simple numerical quadratic technique:

$$A_k^{(j)} = 0 \quad \text{for } k \text{ odd},$$
$$\text{(16)}$$

$$A_k^{(j)} = \frac{1}{2q + 1} \left[y_{\langle j \rangle, 0} + \sum_{l=1}^{q} (y_{\langle j \rangle, l} + y_{\langle j \rangle, -l}) \cos \frac{\pi k l}{2q + 1} \right] \quad \text{for } k \text{ even},$$
$$\text{(17)}$$

$$B_k^{(j)} = 0 \quad \text{for } k \text{ even},$$
$$\text{(18)}$$

$$B_k^{(j)} = \frac{1}{2q + 1} \left[\sum_{l=1}^{q} (y_{\langle j \rangle, l} - y_{\langle j \rangle, -l}) \sin \frac{\pi k l}{2q + 1} \right] \quad \text{for } k \text{ odd},$$
$$\text{(19)}$$

where $y_{\langle j \rangle}$ replace y_j for notational purposes and the number of sampling points is presented for convenience as $r = 2q + 1$ (q – an integer).

The number of sampling points r that must be taken can be related to the maximum frequency ω_{max} of the frequency set using the Nyquist criterion:

$$r \geq N\omega_{max} + 1,$$
$$\text{(20)}$$

where N is the maximum number of the Fourier coefficients that may be retained in calculating the partial variances without interfences between the assigned frequencies. In general, the interference between the higher harmonics will be eliminated when

$$N < \omega_{min} - 1.$$
$$\text{(21)}$$

Larger values of r are numerically desirable for reasons of accuracy, although smaller values are desirable for reasons of computing economy. The minimum value $N = 2$ is usually sufficient in practice because the magnitude of the higher-order terms in the Fourier series tends to decrease rapidly. The final working equations for partial variances and coupled partial variances then become:

$$S_{j/i} = \frac{2}{\sigma_j^2} [(B_{1\omega_i}^{(j)})^2 + (A_{2\omega_i}^{(j)})^2],$$
$$\text{(22)}$$

$$S_{j/il} = \frac{2}{\sigma_j^2} [(A_{\omega_i + \omega_l}^{(j)})^2 + (A_{2(\omega_i + \omega_l)}^{(j)})^2 + (B_{2\omega_i + \omega_l}^{(j)})^2 + (B_{\omega_i + 2\omega_l}^{(j)})^2].$$
$$\text{(23)}$$

The selection of a frequency set is made in accordance with the number of input parameters to be analyzed m with the aid of the recursive algorithm described by Cukier et al. (1978):

$$\omega_1 = \Omega_m, \qquad \omega_i = \omega_{i-1} + d_{m+1-i}, \quad i = 2, \dots, m.$$
$$\text{(24)}$$

Values of the coefficients Ω_m and d_m assumed in the present implementation after

TABLE I

Parameters used in calculating frequency set free of interferences to fourth order (after McRae
et al., 1982) and minimum number of model solutions required by FAST

Number of input parameters m	Ω_m	d_m	Number of model solutions r
1	0	4	–
2	3	8	15
3	1	6	27
4	5	10	47
5	11	20	79
6	1	22	99
7	17	32	175
8	23	40	251
9	19	38	323
10	25	26	411
11	41	56	495
12	31	62	587
13	23	46	695
14	87	76	915
15	87	96	1027

McRae et al. (1982) are listed in Table I together with the minimum number of model
solutions r required for calculation of partial variances. The number r determined by
condition (20) with $N = 2$ depends on the assumed frequency set and strongly grows
with the number of input parameters m to be analyzed. However, for small m the number
of model solutions determined in this way may appear to be too low in the presented
application. Namely, if evaluation of output frequency distribution is desired the number
r should be choosen high enough to obtain a set of model solutions sufficiently numerous
for statistical analysis.

The final step in the FAST method implementaion is the choice of the transformation
functions (4) defining the search curve that traverses the input parameter space. The
transformation functions G_i must satisfy the following differential equation

$$\pi(1 - z^2)^{1/2} p_i(G_i) \frac{\mathrm{d}}{\mathrm{d}z} G_i(z) = 1, \quad z = \sin(\omega_i s) . \tag{25}$$

Then the fraction of the arc length along the search curve which lies between the values
x_i and $x_i + \mathrm{d}x_i$ is equal to $p_i(x_i)\,\mathrm{d}x_i$.

This equation can be solved by quadrature for any distribution function p_i:

$$G_i(s) = x_i = \begin{cases} x_i^k - [p_i^k - \sqrt{(p_i^k)^2 + 2a(S - S_k)}]/a, & a \neq 0 \\ x_i^k + (S - S_k)/p_i^k, & a = 0 \end{cases} \tag{26}$$

for $S_k \leq S \leq S_{k+1}$, where x_i^k, p_i^k – discrete values of input parameter x_i and corresponding values of distribution function p_i,

$$a = (p_i^{k+1} - p_i^k)/(x_i^{k+1} - x_i^k), \qquad S = \text{arc sin}(\sin \omega_i s),$$

$$S_{k+1} = S_k + (p_i^{k+1} + p_i^k)(x_i^{k+1} - x_i^k)/2, \qquad S_1 = -0.5.$$

In many applications the exact form of input parameter distribution is not known. With only information about range of uncertainty $d_i = x_i^{max} - x_i^{min}$ a uniform distribution is the default distribution which corresponds to the following transformation function:

$$G_i(s) = \bar{x}_i + Sd_i = x_i^{min}(0.5 - S) + x_i^{max}(0.5 + S), \qquad (27)$$

where $\bar{x}_i = (x_i^{max} + x_i^{min})/2$.

An additional estimation of central tendency x_i^0 allows us to consider a triangular distribution with the transformating function:

$$G_i(S) = \begin{cases} x_i^0 + (x_i^{min} - x_i^0)(1 - \sqrt{1 - d_i(S - S^0)/(x^{min} - x_i^0)}) \\ \quad \text{for} \quad S \leq S^0, \\ x_i^0 + (x_i^{max} - x_i^0)(1 - \sqrt{1 - d_i(S - S^0)/(x^{max} - x_i^0)}) \\ \quad \text{for} \quad S \geq S^0, \end{cases} \qquad (28)$$

where $S^0 = -0.5 + (x_i^0 - x_i^{min})/d_i$.

The FAST algorithm may be now summarized as follows:

(i) frequency assignment to input parameters;

(ii) calculation of input parameters combinations as sampling points from the search curve using a set of transformation functions $x_i = G_i(s)$;

(iii) multiple model solution for provided parameter combinations;

(iv) evaluation of Fourier coefficients (16)–(19);

(v) evaluation of partial variances (22) and coupled partial variances (23); and

(iv) evaluation of output frequency distribution.

4. Computer Program

The developed Fortran program for the FAST method allows one to investigate the sensitivity-uncertainty of an arbitrary mathematical model with several options for parameter distributions. The analysis is performed in three steps as shown in Figure 1a. A name *file* which is common for all data files involved in analysis is specified by the user at the start of calculations. The file name extensions (*.r*, *.p*, *.m*) are added automatically by the program.

File.d provided by user contains information about input parameter distributions. The available options include the following distributions:

(0) bivariate (x_i^{min}, x_i^{max}),

(1) uniform (x_i^{min}, x_i^{max}),

(2) symmetrical triangle (x_i^{min}, x_i^{max}),

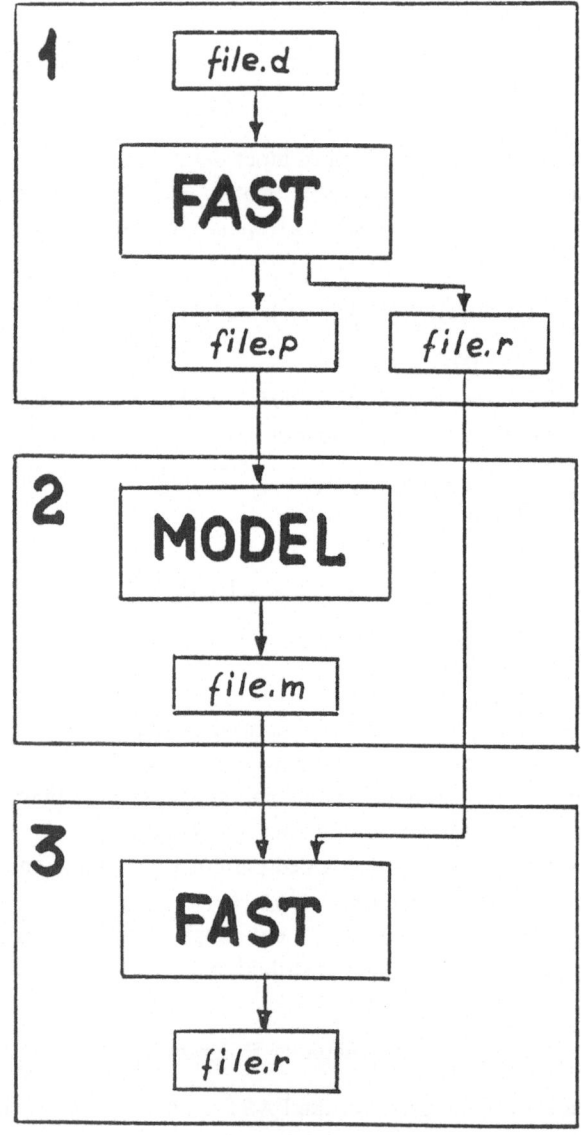

Fig. 1a.

Fig. 1.   Scheme of FAST analysis and the used data files for (a) three-steps option, (b) one-step
option.

(3) asymmetrical triangle $(x_i^{min}, x_i^0, x_i^{max})$,

(4) arbitrary $(x_i^k, p_i^k, k = 1, K, K$ – number of coordinate pairs), and

(5) histogram $(x_i^{min}, x_i^{max}, p_i^k, k = 1, K, K$ – number of classes).

In parenthesis, input data needed to describe parameter frequency distributions are
given.

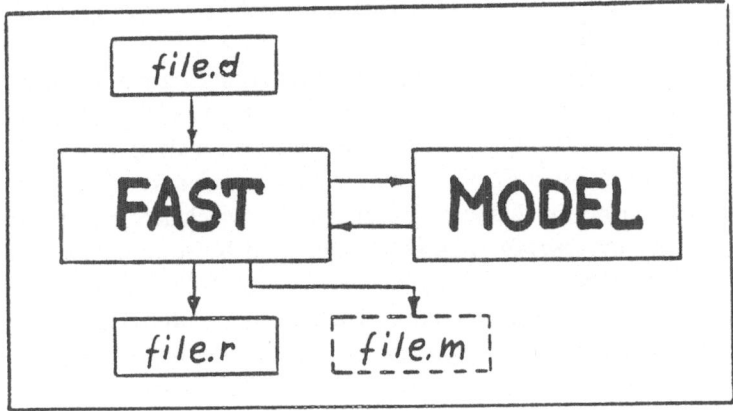

Fig. 1b.

The '0' option is reserved for Walsh Amplitude Sensitivity Procedure (WASP) (Pierce and Cukier, 1982). This is another sensitivity-uncertainty analysis method developed for discrete models where parameter variation is intrinsically two valued (bivariate). The discrete Walsh function series are used instead of the Fourier series but the analysis results are obtained in the same terms as in FAST. The WASP method is exact for continuous models where two valued parameter variations are sufficient (FAST is only an approximate method). This method is ideally suited for the structural analysis of the model to investigate effects of different approximations, e.g., neglecting some terms in the model equations.

The options 0, 1, and 2 can be applied when information on the uncertainty range of a given parameter is only available. The exact analytical solution of Equation (25) determining transformation functions G_i is possible only for very simple shapes of parameter distributions (e.g., uniform or triangular) and, in a general case, the equation is solved numerically. For this purpose the option 4 is designed where the input parameter distribution is represented in an approximate form by a set of coordinates (x_i^k, p_i^k). Using this option the program can be easily extended to include the arbitrary parameter distribution provided in the analytical form by user (e.q., the log-normal distribution which is exceedingly important in air pollution studies). In some case it is convenient to define parameter distributions in the form of frequency histograms (option 5).

In the first step of analysis the FAST program is run to generate input parameter combinations that are written in *file.p*. Information on number of input parameters, number of model state variables and description of parameter distributions are recorded in *file.r*.

File.p contains input data for the second step of computations which are independent of the FAST program and consists in repetitive model solutions. This step results in *file.m* where input parameter combinations and corresponding model solutions are written. This file can then be a subject of statistical analysis to evaluate frequency distributions of model outputs.

In the third step of analysis the FAST program is called again in order to perform Fourier analysis of model solutions and to evaluate partial variances. Information from *file . m* and *file . r* is required as input and the final results of analysis are recorded in *file . r*.

The described three-step FAST analysis is recommended for mathematical models with high requirements for computer time and models which need another data files for their solution. It is the case of long range transport models. However, the FAST analysis may also be performed very simple in one step (Figure 1b) if a model to be analyzed is prepared in the form of a subroutine. The subroutine MODEL calculates output values for a given parameter combination provided by the FAST program. The same input file (*file . d*) must be prepared by user and the results are written in *file . r*. Optionally *file . m* containing model solutions may be created.

5. Application

The FAST method was applied to routine calculations of S transport in Europe using the EMEP-W model (Eliassen and Saltbones, 1975, 1983; Eliassen, 1978) for 1980 meteorological and emission conditions. The EMEP model is a Lagrangian type model where trajectories are first calculated on the base of wind fields. Then the two mass-conservation differential equations are solved on these trajectories to compute SO_2 and $SO_4^=$ concentrations at the receptor points. Finally, dry, wet and total deposition of S are determined with the aid of algebraic equations. The same model version is used as the one considered by Alcamo and Bartnicki (1987).

The present analysis is limited to investigating model error due to uncertainty of parameters which can be treated as constant in time or space. These input parameters of the model and their nominal values assumed in the computations are listed in Table II. Uncertainties in the forcing functions of the model – wind, precipitation, and emission field – were not included into the analysis because the FAST method is not well suited to handle with parameters which inherently vary in time and space. For this purpose, the variational method (Uliasz, 1983, 1985) can be recommended. The five model outputs were taken into account – 1-yr averaged values of SO_2 concentration (C_{SO_2}), $SO_4^=$ concentration (C_{SO_4}), dry (D_d), wet (D_w), and total (D_t) deposition of S. Calculations were carried out for various shape and range of parameter frequency distributions and three source-receptor combinations: (1) German Democratic Republic (GDR) – Illmitz (Austria), (2) United Kingdom (UK) – Rörvik (Southern Sweden), and (3) Netherlands (NL) – Tange (Denmark). The same frequency distributions were assumed for all input parameters in a given run of analysis what is clearly a great simplification of the real long range transport situation. However, it should be pointed out that it is not a limitation of the methodology since the FAST program allows us to apply arbitrary frequency distribution for individual parameters if this information is available. Every run of the FAST analysis required 411 model solutions and was performed using the three-steps algorithm (Figure 1a).

TABLE II

Input parameters of the EMEP long range transport and their nominal values

Notation	Explanation	Value	Unit
v_d	Deposition velocity for SO_2	8×10^{-3}	$m\,s^{-1}$
v_{ds}	Deposition velocity for $SO_4^=$	2×10^{-3}	$m\,s^{-1}$
h	Mixing height	1000	m
k_t	Transformation rate of SO_2 to $SO_4^=$	2×10^{-6}	s^{-1}
k_w	Wet deposition of SO_2 (nonzero when it is raining)	3×10^{-5}	s^{-1}
α	Additional dry deposition in the same grid where emission occurs	0.15	–
β	Part of S emission assumed to be emitted directly as sulphate	0.05	–
$\varkappa$	Overall decay rate for $SO_4^=$	4×10^{-6}	s^{-1}
a	Transfer coefficient used in calculation of S concentration in precipitation from $SO_4^=$	0.69	–
b	Background concentration of S in precipitation	0.0	$mg\,S\,L^{-1}$

The results obtained from the FAST analysis for total S deposition D_t are summarized in Table III. The model uncertainty is characterized by a coefficient of variation c_v which varies with uncertainty range of input parameters. The model sensitivity to uncertainties in the individual input parameters is expressed by partial variances (Equation (12)). They are presented here as percentages of the total variance of the model output, e.g., the value 38% given in the first row of Table III for the partial variance related to mixing height h means that 38% of the total S deposition variance is caused by uncertainties of mixing height. The partial variances do not change significantly for various shapes of parameter frequency distributions. Only the partial variances evaluated by the WASP method for bivariate parameter distributions differ slightly from these calculated by FAST. The model sensitivity is then practically independent of the assumed shape of parameter frequency distributions.

The partial variances also vary very little with the uncertainty range of input parameter distributions (see triangular distribution ± 20, ± 50, and $\pm 80\%$ in Table III). However, in the presented calculations the frequency distributions were changed exactly in the same manner for every input parameter. It is obvious that the model sensitivity may be significantly redistributed if, e.g., the uncertainty range of only one or some selected parameters are changed.

The sums of partial variances shown in Table III are less than 100%. The rest of the total variance is related to uncertainty coupled among groups of input parameters and may be expressed by coupled partial variances. However, in the case of the EMEP model, the partial variances presented can be used as a sufficiently accurate measure of the model sensitivity since they explain about 90% of the total variance of model results. The effect of coupled uncertainties increases with the growth of uncertainty range of input parameters.

TABLE III

Mean value, coefficient of variation c_v and partial variances of total S deposition D_t for the performed runs of FAST analysis (results of the WASP (*) and the linear analysis (**) are added for comparison)

Distribution of input uncertainty	D_t (mg m^{-2} yr^{-1})	c_v (%)	Partial variances (%)										
			v_d	v_{ds}	h	k_t	k_w	α	β	$\varkappa$	a	b	Σ
					GDR–Illmitz.								
Triangle $\pm 20\%$	585	6.8	0.0	0.9	38.0	10.7	8.2	3.7	0.6	5.8	25.0	0.0	92.9
Triangle $\pm 50\%$	594	17.4	0.0	0.8	37.1	10.4	7.1	3.5	0.6	5.6	24.3	0.2	90.4
Triangle $\pm 80\%$	614	30.0	0.3	0.8	35.0	9.7	7.0	3.0	0.7	5.2	22.6	0.4	84.3
Asymet. triangle -50, $+100\%$	557	23.8	0.1	0.8	35.3	9.0	7.4	4.6	0.7	7.2	24.4	0.1	89.6
Uniform $\pm 50\%$	606	25.4	0.1	1.1	37.6	11.0	7.8	3.4	0.7	5.8	25.3	0.1	92.9
Irregular $\pm 50\%$	607	26.2	0.2	1.1	37.6	11.2	7.7	3.4	0.8	5.8	25.6	0.0	93.4
Bivariate $\pm 50\%$ * (WASP)	656	49.1	1.2	1.1	29.7	11.3	5.5	2.6	1.0	5.3	24.4	0.0	82.1
Linear analysis ** $\pm 50\%$	583	–	0.6	4.5	27.0	13.4	12.5	7.8	3.3	10.7	20.3	–	–
					NL–Tange								
Triangle $\pm 50\%$	57	24.0	7.8	0.8	16.0	15.8	2.5	1.6	1.0	8.3	36.3	0.1	89.6
					UK–Rörvik								
Triangle $\pm 50\%$	342	26.0	6.4	0.1	17.9	15.9	3.4	1.3	1.4	6.7	36.3	0.1	89.5

The total deposition for the GDR–Illmitz calculation is the most sensitive to mixing height h ($\approx 37\%$), followed by the transfer coefficient a ($\approx 25\%$), the transformation rate of SO_2 to $SO_4^=$ k_t ($\approx 10\%$), wet deposition rate of SO_2 k_w ($\approx 8\%$), and overall decay rate for $SO_4^=$ $\varkappa$ ($\approx 6\%$). It is interesting to note that for the other source-receptor combinations: UK–Rörvik and NL–Tange the model sensitivity is quite different. The highest partial variances are related to coefficient a ($\approx 36\%$), h ($\approx 17\%$), k ($\approx 16\%$), and $\varkappa$ ($\approx 7\%$). The partial variances for deposition velocity v_d are about 7% while for the case of GDR–Illmitz they are practically equal to zero. The differences of model sensitivity between various source-receptor combinations are even better visible for another model outputs (Figure 2). So, the model sensitivity may depend strongly on meteorological and emission conditions which are specific for a given source-receptor pair. The concentrations and S depositions at Illmitz are significantly higher than for the two remaining source-receptor calculations.

Since the S background concentration in precipitation b is assumed to be zero the model output is nonsensitive in relation to this parameter and all corresponding partial variances should be exactly equal to zero. The small nonzero values of these partial variances in Table III indicate an approximate character of the FAST method.

In Table III, the results of linear sensitivity analysis are also given. The first order

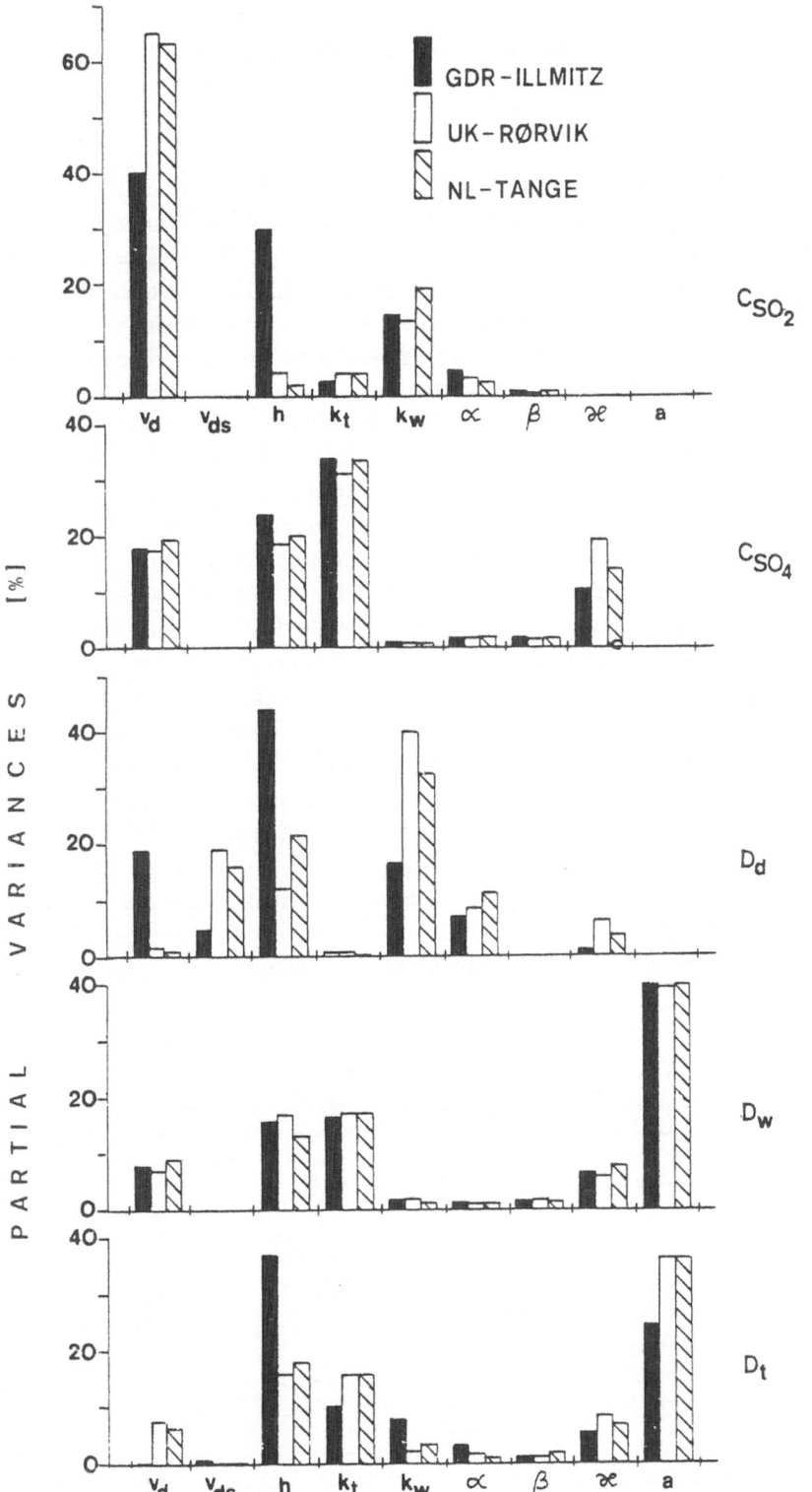

Fig. 2. Partial variances of model outputs for three source-receptor combinations and triangular $\pm 50\%$ frequency distribution of input parameters.

normalized sensitivity coefficients ρ_i are calculated as follows

$$\rho_i = |\Delta F_i| / \sum_{l=1}^{m} |\Delta F_l|, \tag{29}$$

where ΔF_i is a change of model output caused by perturbation Δx_i of ith input parameter x_i while all other parameters are fixed to their nominal values. The results of the linear sensitivity analysis are in general not equivalent to partial variances which describe effects of simultaneous parturbations of all parameters. One can expect that for decreasing input uncertainty ranges the differences between them are also decreasing.

Finally, the FAST method is compared against the Monte-Carlo approach that was applied to the EMEP-W model in the same manner as described by Alcamo and Bartnicki (1987). The number $r = 1500$ of model solutions was assumed to be sufficiently high to create a statistically significant sample in all considered cases. During the computations performed by Alcamo and Bartnicki (1987), the output frequency distributions from the present and previous model runs were compared. They called a sample statistically significant, if the difference between clases in current and previous runs was less than 1%. A minimum of 400 and maximum of 1500 runs were required in their analysis. The number of model runs required by the FAST depends on the number of input parameters involved in the analysis (Table IV). The performed FAST analysis with 10 input parameters needs $r = 411$ model solutions, but this number may be reduced to $r = 323$ since $b = 0$ and uncertainties of 9 parameters only are taken into account. Thus, the FAST method is more efficient than the Monte-Carlo approach. Figure 3 shows that the frequency histograms of the two model outputs (C_{SO_2} and D_t) obtained by the FAST and Monte-Carlo methods are very close. The mean values and coefficients of variation are practically the same in both cases. Some differences occur in the higher statistical moments of frequency distributions and consequently in skewness and kurtosis (Table IV). However, for practical applications one can assume that the FAST method is able to provide the same model output frequency distribution as the Monte-Carlo approach.

TABLE IV

Comparison of frequency distributions of model output obtained by FAST and Monte-Carlo analysis (GDR–Illmitz, triangular ±50% distribution of input parameters)

Model output	Method	Mean	$c_v(\%)$	Skewness	Kurtosis
C_{SO_2} ($\mu g\,m^{-3}$)	FAST	1.126	17.53	0.850	1.151
	Monte-Carlo	1.127	17.66	0.738	0.648
	$\Delta(\%)$	0.1	0.7	13.2	43.7
D_t ($g\,m^{-2}\,yr^{-1}$)	FAST	0.594	17.43	0.600	0.385
	Monte-Carlo	0.596	17.42	0.672	0.541
	$\Delta(\%)$	0.2	0.1	12.0	40.5

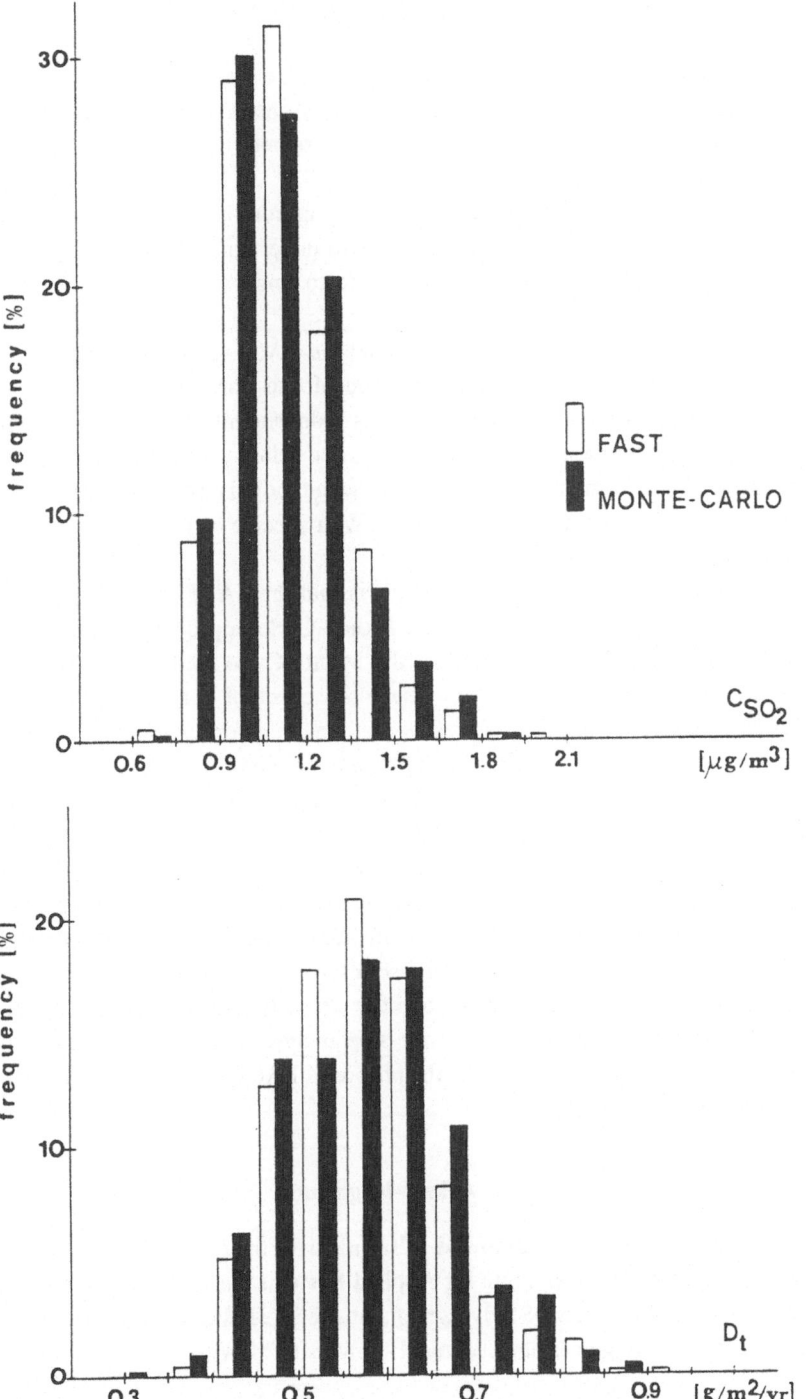

Fig. 3. Frequency histograms of SO₂ concentration (C_{SO_2}) and total deposition D_t calculated by FAST and Monte-Carlo method (GDR–Illmitz), triangular $\pm 50\%$ frequency distribution of input parameters.

6. Conclusions

(1) The FAST method allows evaluation of the model uncertainty due to uncertainties in input parameters in the same manner as the Monte-Carlo analysis. The frequency distributions of the EMEP model outputs obtained by FAST and Monte-Carlo approach are equivalent.

(2) The FAST method provides additionally in comparison with the Monte-Carlo analysis information on the model sensitivity to uncertainties in individual parameters in the form of partial variances. It helps one to understand what drives the model uncertainty.

(3) In the application to the EMEP model the FAST analysis appears to be 2 to 4 times more computationally efficient than the Monte-Carlo analysis.

(4) The sensitivity of the EMEP model is independent of the assumed shape and range of parameter frequency distributions. The relative contributions to the model uncertainty arising from the uncertainties of the individual input parameters vary significantly for the different source-receptor pairs with specific emission and meteorological conditions.

The main conclusion of the present study is that the FAST method is more efficient and provides more information than the Monte-Carlo approach in application to the EMEP model. Thus FAST can replace the Monte-Carlo method in evaluation of the model uncertainties in all problems where the effect of correlations between input parameters can be neglected. The present mathematical formulation of FAST does not allow one to take into account correlated parameters. The FAST analysis is, in principal, limited to input parameters constant in time and space. Distributed parameters like the emission and wind fields in the EMEP model can be taken into account under assumption that their relative perturbations are indepenent of time and space. It can be achieved by introducing some scaling parameters for these fields (Uliasz, 1985).

The computer program developed specially for the sensitivity-uncertainty analysis of the EMEP long range transport model also allows one to perform similar analysis for other mathematical models. Two options for interaction between the analyzed model and the FAST program (three-steps or one-step analysis) and several options to describe uncertainty of input parameters make the proposed FAST program applicable to a wide class of models.

Acknowledgments

The author is grateful to J. Alcamo and J. Bartnicki for their help and discussions during his work in International Institute for Applied Systems Analysis. The above work was performed under Contracted Study Agreement 1986/1987 between IIASA and Institute of Environmental Engineering, Technical University of Warsaw.

References

Alcamo, J. and Bartnicki, J.: 1985, *An Approach to Uncertainty of a Long Range Air Pollutant Transport Model*, IIASA Working Paper, RWP-85–88.

Alcamo, J. and Bartnicki, J.: 1987, *Atmos. Envir.* **21**, 2121.

Cukier, R. I., Levin, H. B., and Shuler, K. E.: 1978, *J. Compt. Phys.* **26**, 1.

Eliassen, A.: 1978, *Atmos. Envir.* **12**, 479.

Eliassen, A. and Saltbones, J.: 1975, *Atmos. Envir.* **9**, 425.

Eliassen, A. and Saltbones, J.: 1983, *Atmos. Envir.* **17**, 1457.

Falls, A. H., McRae, G. J., and Seinfeld, J. H.: 1979, *Int. J. Chem. Kinetics* **XI**, 1137.

Koda, M., Dogru, A. H., and Seinfeld, J. H.: 1979, *J. Compt. Phys.* **30**, 259.

McRae, G. J., Tilden, J. W., and Seinfeld, J. H.: 1982, *Computer Chem. Eng.* **6**, 15.

Pierce, T. H., Cukier, R. I., and Dye, J. L.: 1981, *Math. Biosciences* **56**, 175.

Pierce, T. H. and Cukier, R. I.: 1982, *J. Compt. Phys.* **41**, 427.

Tilden, J. W. and Seinfeld, J. H.: 1982, *Atmos. Envir.* **16**, 357.

Uliasz, M.: 1983, *Z. Meteorol.* **33**, 355.

Uliasz, M.: 1985, *Z. Meteorol.* **35**, 340.

APPLICATION OF THE 'FAST'-METHOD TO A LONG TERM INTERREGIONAL AIR POLLUTION MODEL

W. KLUG and B. ERBSHÄUßER

Technische Hochschule Darmstadt, Institut für Meteorologie, D 6100 Darmstadt, F.R.G.

(Received November 17, 1987; revised May 24, 1988)

Abstract. The Fourier Amplitude Sensitivity Test (FAST) is applied to an interregional Air Pollution Model which simulates SO_2/SO_4-concentrations and -depositions as an annual average. The results are discussed and problems connected with the application of the FAST-method are reported.

1. Introduction

In understanding the behavior of a mathematical model of a physical system it is necessary to determine the sensitivity of the solution to variations of input parameters. It is important to know how sensitive the output variables react to changes of, or uncertainties in, the parameters, and which of the variables are sensitive to which parameters. One possible way for model examination is to solve the model equations over and over again, varying one parameter at a time over a series of values, while holding all other parameters fixed. But this attempt soon becomes prohibitive in time and produces an enormous number of results, so that the analysis would be a major problem. Other ways of estimating the sensitivity are the 'direct' or the 'indirect' ('brute-force') method. By varying parameters one by one one can get partial derivatives as 'sensitivity coefficients'. But these test methods are limited for only small perturbations in input parameters in the vicinity of their unperturbed values. In addition, the 'sensitivity coefficients' calculated by the 'brute-force' method are associated with large errors. The 'Fourier Amplitude Sensitivity Test' (FAST) is nonlinear. This allows us to examine larger deviations from the nominal parameter values. Furthermore, all parameters can be varied simultaneously.

The subject of this paper is to report on the application of the 'FAST'-method to study the sensitivity of a simple Long Term Interregional Air Pollution Model. This model will be described in detail in Section 2. After a few remarks on the conditions of implementation in Section 3 the results will be demonstrated and discussed shortly.

Finally we state briefly several problems which occur during the course of the 'FAST'-analysis.

2. The Long Term Interregional Air Pollution Model

The model to be examined estimates the mean annual long range transport of S in the atmosphere over Europe. It is a two-dimensional Eulerian box model which assumes immediate mixing in the horizontal and vertical. Each box covers (Figure 1) an area of

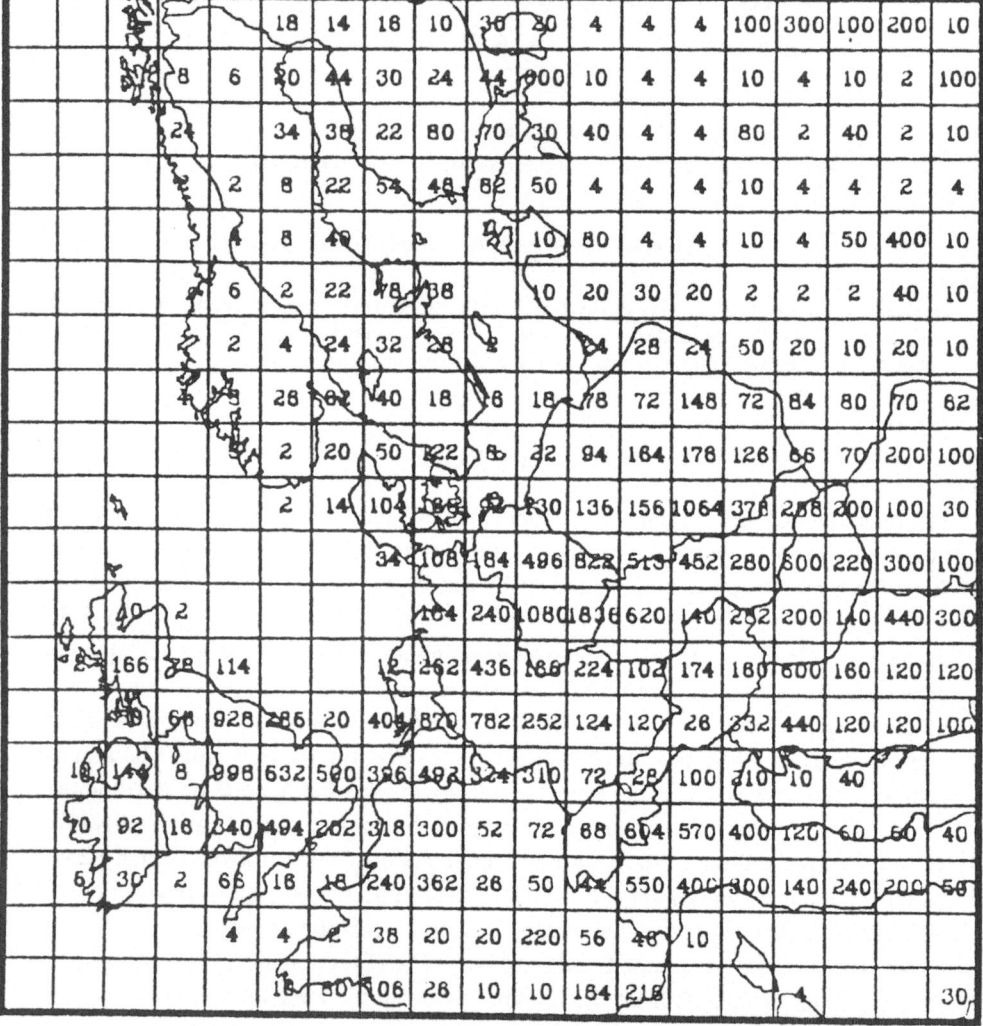

Fig. 1.   The (19, 19) Emission Data Base; grid length 150 km; taken from EMEP (Dovland and Saltbones, 1986).

150×150 km^2 and has a constant height (mixing height) of 1000 m. The concentration within each box is constant (Klug, 1982; Klug and Lüpkes, 1985). The S emission inventory of the year 1978 is used, taken from the 'Co-operative Program for Monitoring and Evaluating of the Long Range Transmission of Air Pollutants in Europe' (EMEP) (3).

The temporal concentration changes for SO$_2$ and SO$_4$ are caused by the source term, an advection term and a decay term as follows:

for SO$_2$:

$$\frac{\delta S_2}{\delta t} = (1 - \alpha)\,\frac{Q}{h} - \mathbf{v} \cdot \nabla S_2 - \left(\frac{v_{d2}}{h} + k_c + k_{r2}\right);$$

TABLE I

Precipitation wind rose

Sector	Probability of wind direction	Wind velocity $(m\ s^{-1})$	Probability of precipitation
0°– 45°	0.105	8.5	0.07
45°– 90°	0.090	8.5	0.09
90°–135°	0.070	8.1	0.09
135°–180°	0.085	9.3	0.05
180°–225°	0.160	11.0	0.17
225°–270°	0.210	11.1	0.16
270°–315°	0.165	10.2	0.10
315°–360°	0.115	9.0	0.11

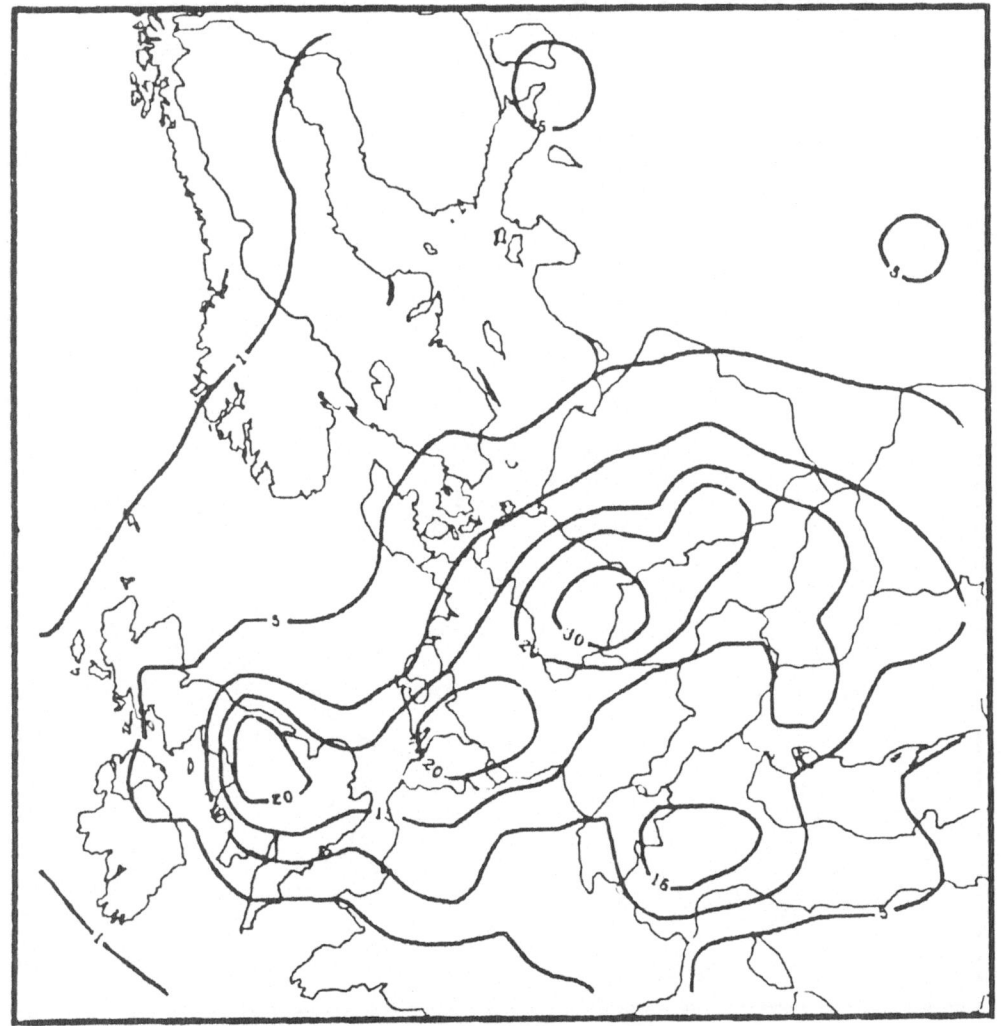

Fig. 2.   Concentrations of SO_2 in ($\mu g\ SO_2\ m^{-3}$).

TABLE II

Parameter values used in the model

Washout rate	$w_r = 1 \times 10^{-4}$ (s^{-1})
SO_2-deposition velocity	$v_{d2} = 8 \times 10^{-3}$ $(m\,s^{-1})$
SO_4-deposition velocity	$v_{d4} = 2 \times 10^{-3}$ $(m\,s^{-1})$
SO_2–SO_4 conversion rate	$k_c = 2.8 \times 10^{-6}\,(s^{-1})$
Fraction of SO_2 emitted directly as SO_4	$= 0.05$

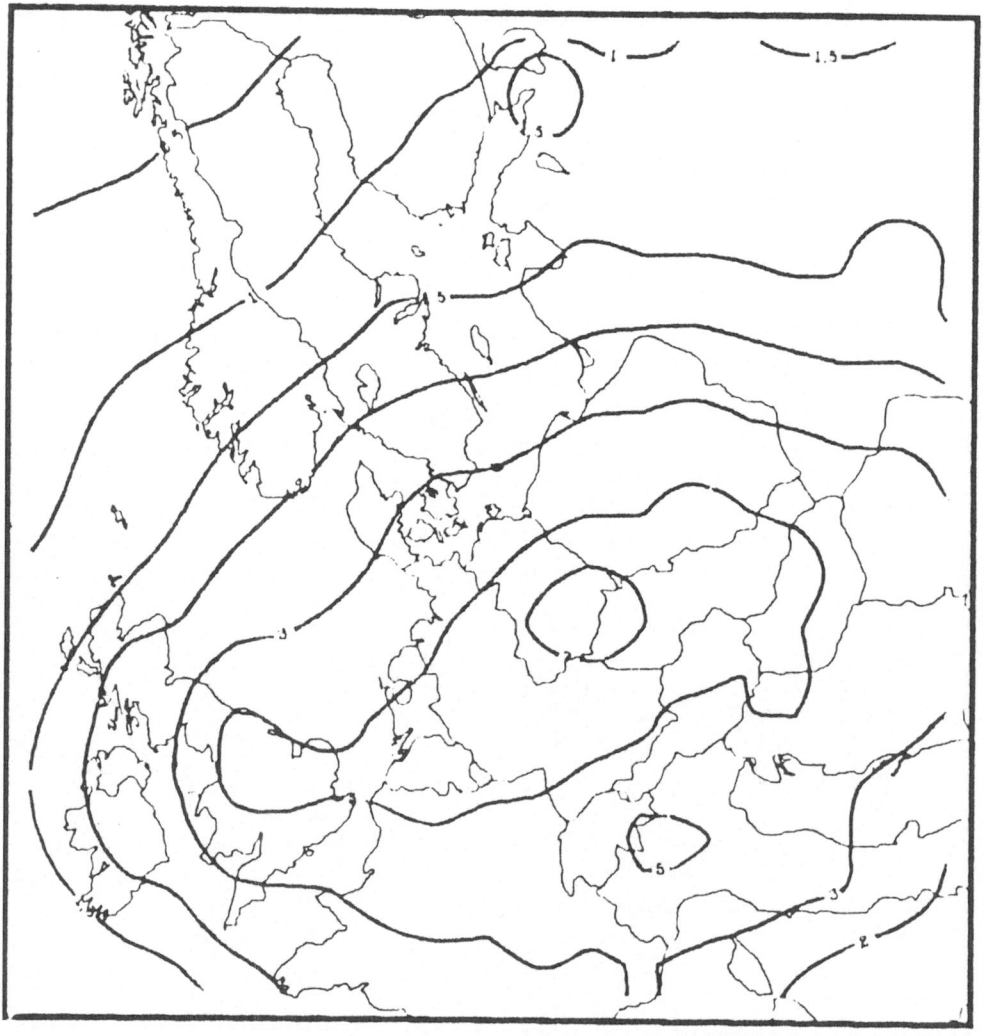

Fig. 3.   Concentrations of SO_4 in ($\mu g\, SO_4\, m^{-3}$).

for SO_4:

$$\frac{\delta S_4}{\delta t} = \frac{3}{2}\left(\alpha\,\frac{Q}{h} + k_c S_2\right) - \mathbf{v}\cdot\nabla S_4 - \left(\frac{v_{d4}}{h} + k_{r4}\right)S_4,$$

with Q the source strength, the fraction of SO_2 emitted directly as sulfate, v_{d2} and v_{d4} the dry deposition velocities for SO_2 and SO_4, k_c the chemical conversion rate, k_{r2} and k_{r4} the wet deposition rates multiplied with the average annual precipitation probability of 0.11.

For a description of the windfield and the wet decay rates the model makes use of a precipitation wind direction rose, divided into eight equal sectors. Each sector is

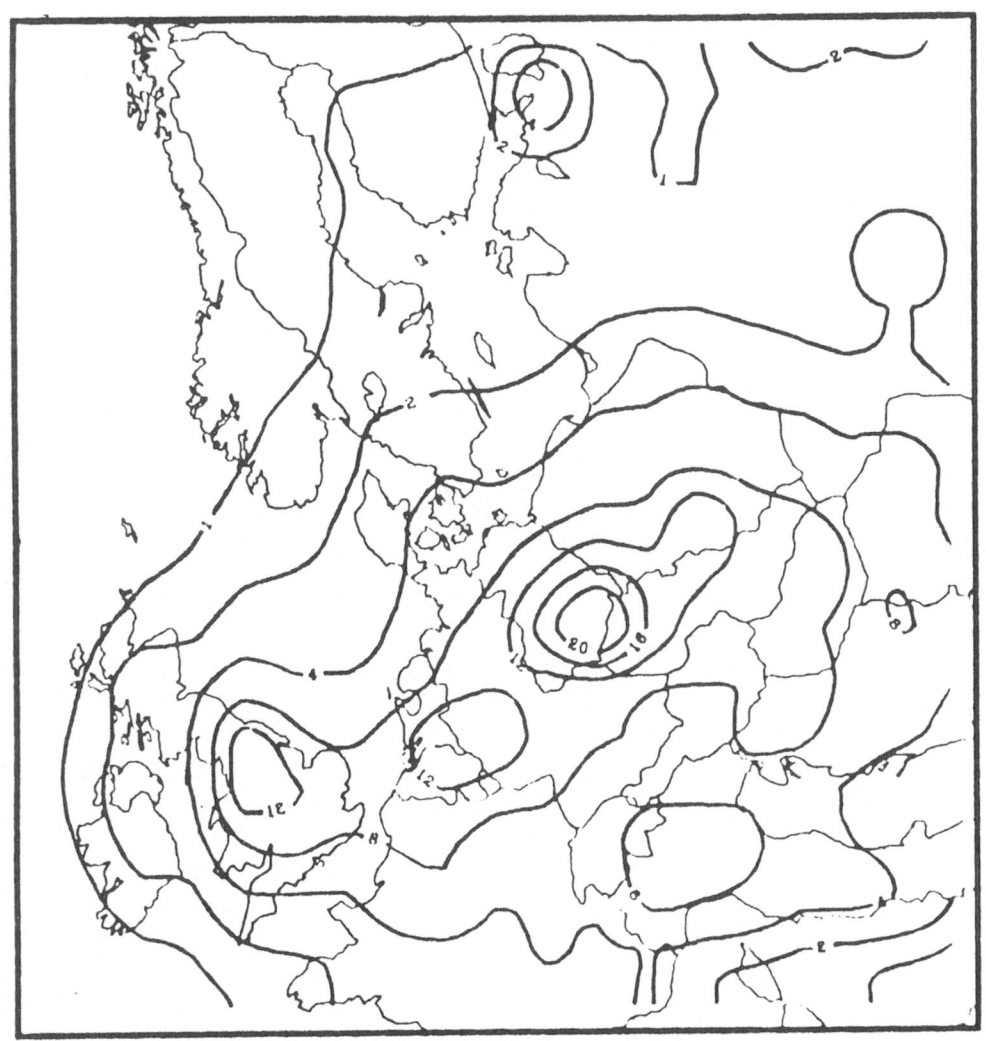

Fig. 4. Total (wet + dry) deposition of SO_2 in (g SO_2 m^{-2} yr^{-1}).

characterized by an average wind velocity, a probability of wind direction and a probability of precipitation (Table I). Wind velocity v and the decay rates k_{r2} and k_{r4} are constant for each sector. The physical and chemical parameters are listed in Table II.

After using the stationarity condition and solving the advection term by an upstream-differencing scheme the SO_2- and SO_4-concentration can easily be obtained:

For SO_2:

$$S_2 = \frac{(1 - \alpha) \dfrac{QL}{h} + |u|\, S_{2,\,-x} + |v|\, S_{2,\,-y}}{|u| + |v| + \left(\dfrac{v_{d2}}{h} + k_c + k_{r2}\right) L} \; ;$$

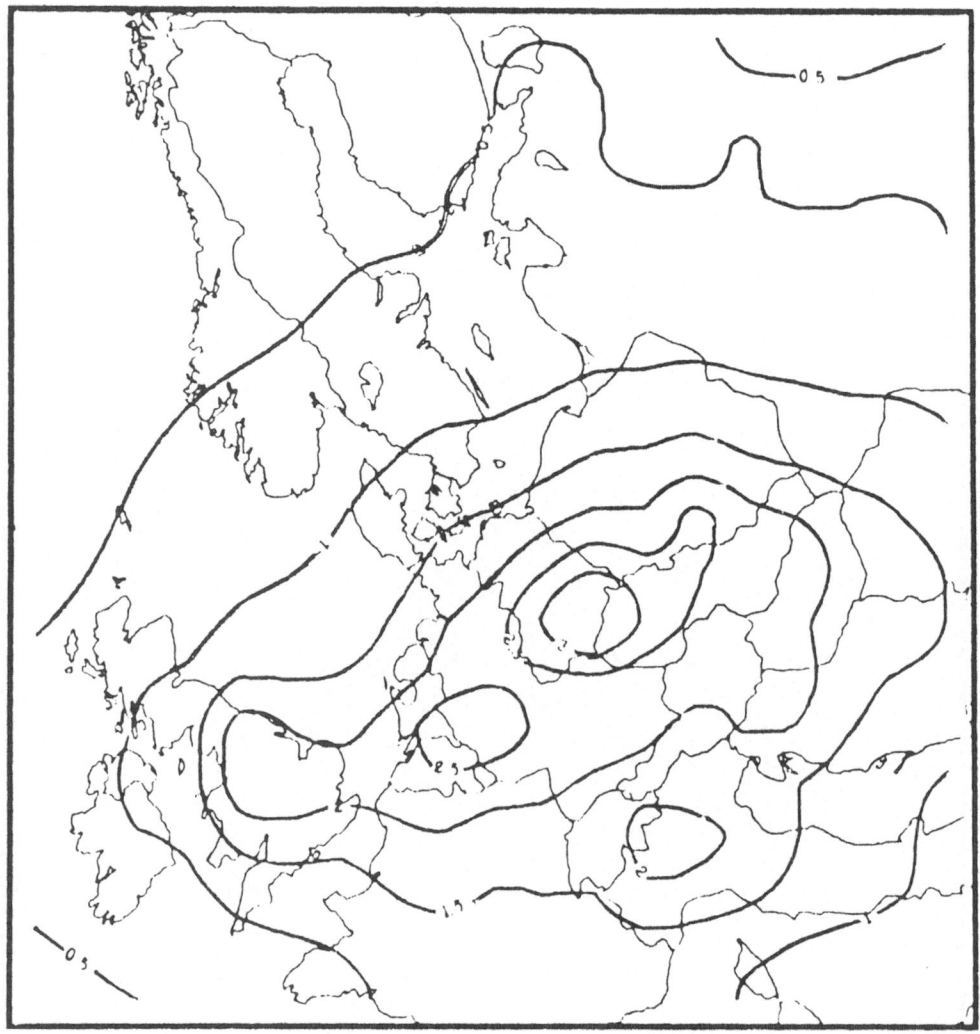

Fig. 5.   Total (wet + dry) deposition of SO_4 in (g SO_4 m^{-2} yr^{-1}).

for SO_4:

$$S_4 = \frac{\frac{3}{2}\left(\alpha\,\frac{Q}{h} + k_c S_2\right)L + |u|\,S_{4,\,-x} + |v|\,S_{4,\,-y}}{|u| + |v| + \left(\frac{v_{d4}}{h} + k_{r4}\right)L}\,,$$

where L describes the box dimension and $S_{i,\,-x}$, respectively, $S_{i,\,-y}$ is the advected concentration on the up-wind side of the box with the Cartesian component of the wind

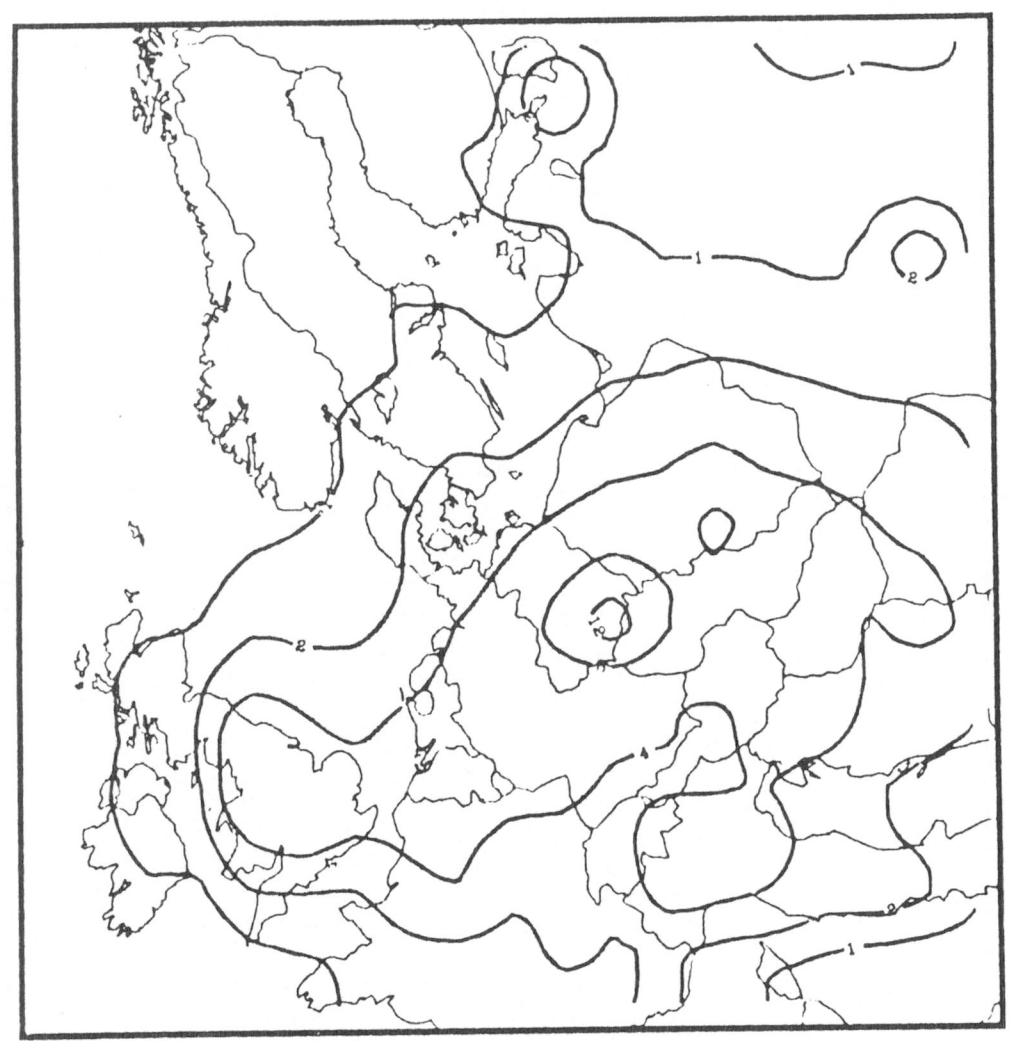

Fig. 6.   Total deposition of S in ($g\,S\,m^{-2}\,yr^{-1}$).

velocity component u, respectively, v. The values of the total deposition (dry plus wet deposition) are calculated by the following equations:

for SO_2:

$$D_{t2} = v_{d2}S_2 + K_{r2}hS_2 ;$$

for SO_4:

$$D_{t4} = v_{d4}S_4 + k_{r4}S_4 .$$

The total S deposition can be computed as

$$D_s = \tfrac{1}{3}D_{t4} + \tfrac{1}{2}D_{t2} .$$

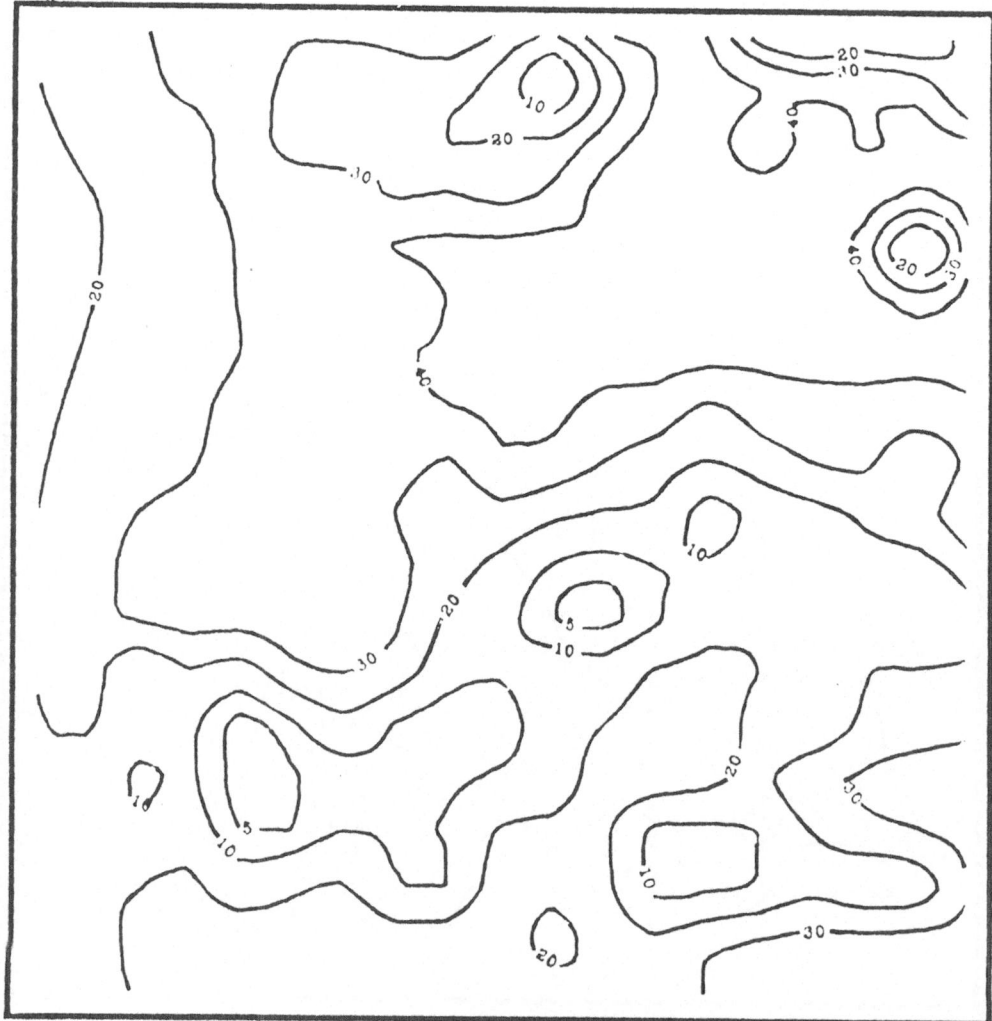

Fig. 7. SO_2-concentration variances in %; parameter: SO_2-deposition velocity.

A final comparison between computed results and observed regional distributions yields better agreement after a windfield modification:

$$\hat{v} = v_k \mathbf{v}; \quad \text{with the modification factor} \quad v_k = 0.8.$$

Figures 2 to 6 show the calculated values of concentrations and total deposition of SO_2, SO_4, and S.

3. The Application of the 'FAST'-Method

The theory of the applied 'Fourier Amplitude Sensitivity Test' was developed by Cukier *et al.* (1975) and is described in detail by Cukier *et al.* (1973, 1975, 1978) and Schaibly

Fig. 8. SO_2-concentration variances in %; parameter: SO_2-washout rate.

and Shuler (1973). The used computational implementation of this test method is found in publications of Uliaz (1986a, b) or McRae *et al.* (1982). Therefore, in this paper we restrict ourselves to state the input parameters. The 'FAST'-method yields partial variances. The partial variance is the fraction of the variance of the output function due to the variation of one special input parameter when the output function is averaged over the variation of all other parameters. So the partial variance is a measure of the sensitivity of the output variable to the variation of one input parameter.

In our Long Term Interregional Air Pollution Model we consider simultaneous perturbations of eight several input parameters (Figures 7 to 14) (deposition velocity of SO_2 and SO_4, washout rate of SO_2 and SO_4, SO_2–SO_4 conversion rate, fraction of

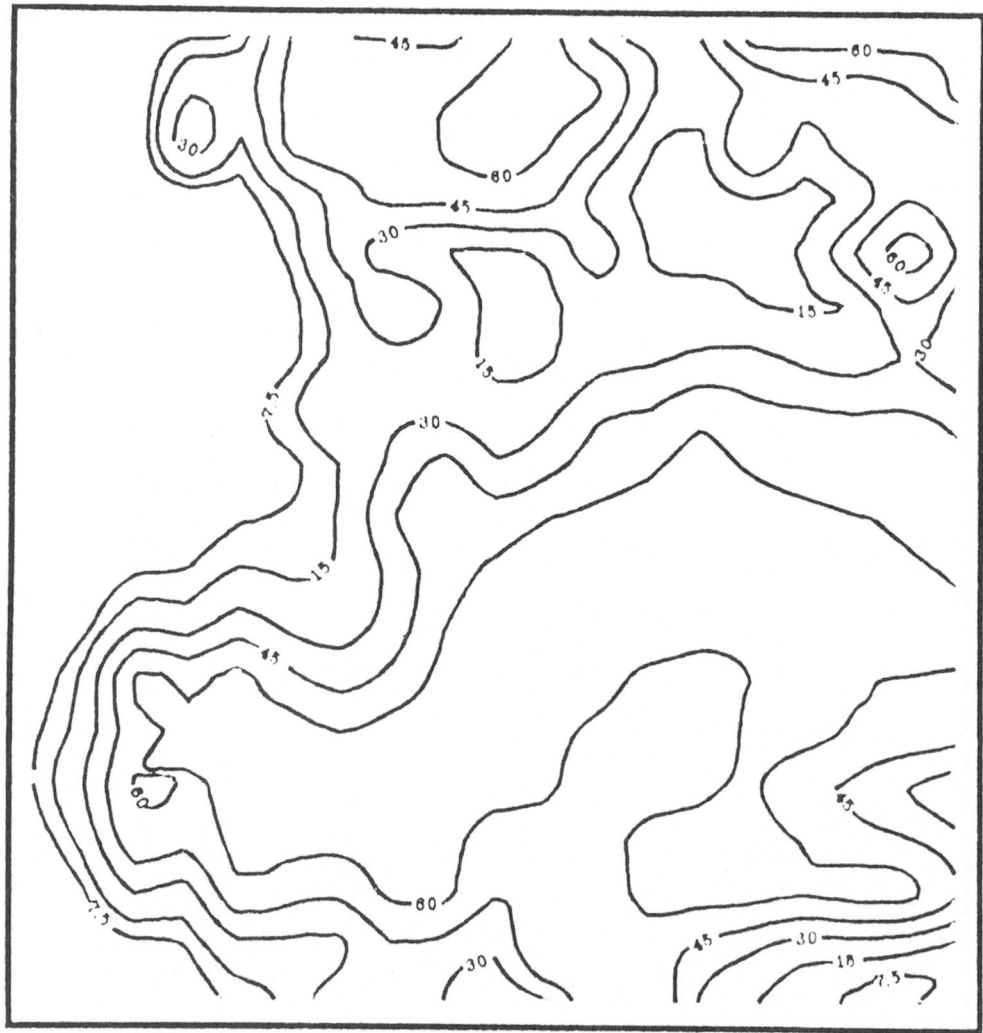

Fig. 9. SO_2-concentration variances in %; parameter: mixing height.

SO_2 emitted directly as SO_4, mixing height, wind velocity). The range of variation of each parameter is limited to $\pm 10\%$. For including the variation of wind velocity an examination of the eight windsectors would be necessary. So we vary the modification factor, and the wind velocity changes in each sector in the range of $\pm 10\%$.

It stands to reason that these parameter perturbations influence the concentration and total deposition distributions of SO_2 and SO_4 as output variables. In the following figures a selection fo the computed partial variances is given. The results can be interpreted as follows:

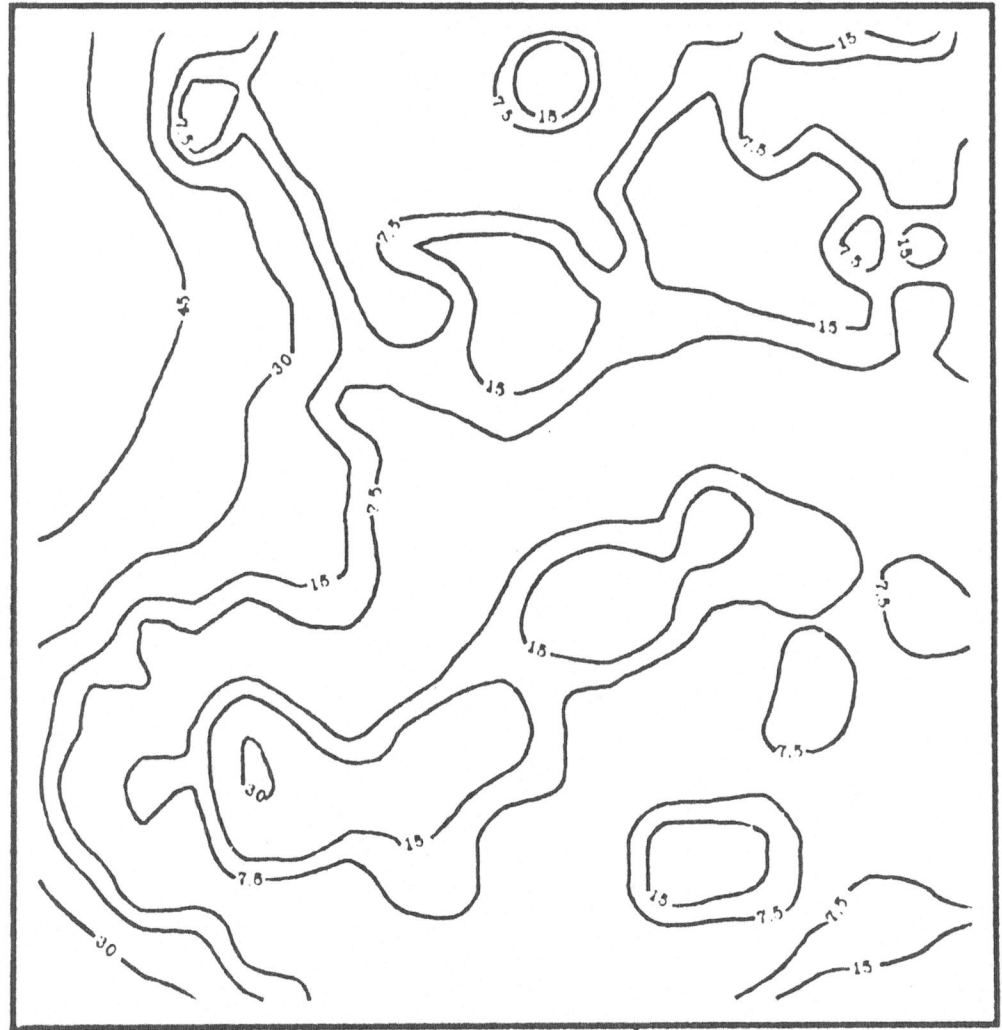

Fig. 10. SO_2-concentration variances in %; parameter: wind velocity.

3.1. SO₂-CONCENTRATION

– In the model equations one can already recognize that the SO_4-decay rates (SO_4 deposition velocity, SO_4 washout rate) have no influence on the SO_2-output.

– The SO_2-decay rates (SO_2-deposition velocity, SO_2-washout rate, SO_2–SO_4 conversion rate) have maximal influence in regions of minimal SO_2-concentration (and *vice versa*). The conversion rate is of little importance.

– The influence on SO_2-concentration caused by the source portion of SO_4 is negligible.

– In the vicinity of high source strength areas the SO_2-concentration is most sensitive to the mixing height. The wind velocity has variable influence. The highest values of sensitivity appear far from the main emissions and also close to them.

Fig. 11. SO_4-concentration variances in %; parameter: SO_4-washout rate.

Close to the large source areas the SO_2-concentration is mainly influenced by the mixing height. Far from the sources the SO_2-concentration is largely determined by the SO_2-decay rates and the wind velocity.

3.2. SO_4-CONCENTRATION

– The decay rates (SO_2-deposition velocity, SO_2-washout rate, SO_4-deposition velocity, SO_4-washout rate) have maximal influence in regions of minimal SO_4-concentration (and *vice versa*).

– The SO_4-washout rate has more influence on the SO_4-concentration than the other deposition rates. The SO_2–SO_4-conversion rate shows the same tendency. As compared with the SO_2-concentration its influence is larger.

Fig. 12. SO_4-concentration variances in %; parameter: SO_2–SO_4 conversion rate.

– The influence of the source portion of SO_4 is low, but close to the main sources it is not negligible.

– In the neighborhood of large source areas the mixing height plays an important part again.

– In regions with high emissions the SO_4-concentration is dominated by the wind velocity.

Close to the large source areas the SO_4-concentration is determined by the mixing height and the wind velocity. Far from the sources the importance of the decay rates increases more and more.

Fig. 13. SO_4-concentration variances in %; parameter: mixing height.

3.3. SO_2-TOTAL DEPOSITION

– The model equations already show that the SO_2-total deposition is independent of the SO_4-decay rates (SO_4-deposition velocity, SO_4-washout rate).

– SO_2-deposition velocity and SO_2-washout rate have their largest influence close to the large sources. Their importance diminishes with increasing distance from the source maximum, but far from the sources it increases again. The partial variances due to SO_2–SO_4-conversion rate show the opposite distribution. The transformation has more influence in regions of minimal partial variances due to SO_2-deposition velocity for SO_2-washout rate.

Fig. 14. SO_4-concentration variances in %; parameter: wind velocity.

– The SO_4 emission portion and the conversion rate have a similar influence. But the influence of the emission part is low again.

– With increasing distance from the large source areas the importance of the mixing height and wind velocity goes through a minimum. In the vicinity of high source areas and again for larger distances both parameters determine the result.

Close to the large source areas the SO_2-total deposition is mainly influenced by wind velocity, but also the SO_2-deposition rates and the mixing height are important. Far away from the sources the SO_2-total deposition is dominated by the wind velocity. In between the SO_2–SO_4-conversion rate becomes the determining parameter.

Fig. 15. SO_2-total deposition variances in %; parameter: SO_2–SO_4-conversion rate.

3.4. SO_4-TOTAL DEPOSITION

The SO_2-deposition rates (SO_2-deposition velocity, SO_2-washout rate) have their greatest influence in regions of minimal SO_4-total deposition. The partial variances increase with increasing distance from the maximal deposition. The sensitivity coefficients due to SO_4-deposition rates (SO_4-deposition-velocity, SO_4-washout rate), however, traverse a region with minimal values. In this domain the SO_4-total deposition becomes especially sensitive to changes in the SO_2–SO_4 conversion rate.

– The variations of SO_4 emission portion play a role in vicinity of the main sources.

– Only at large distances from the main sources the influence of the mixing height is worth mentioning.

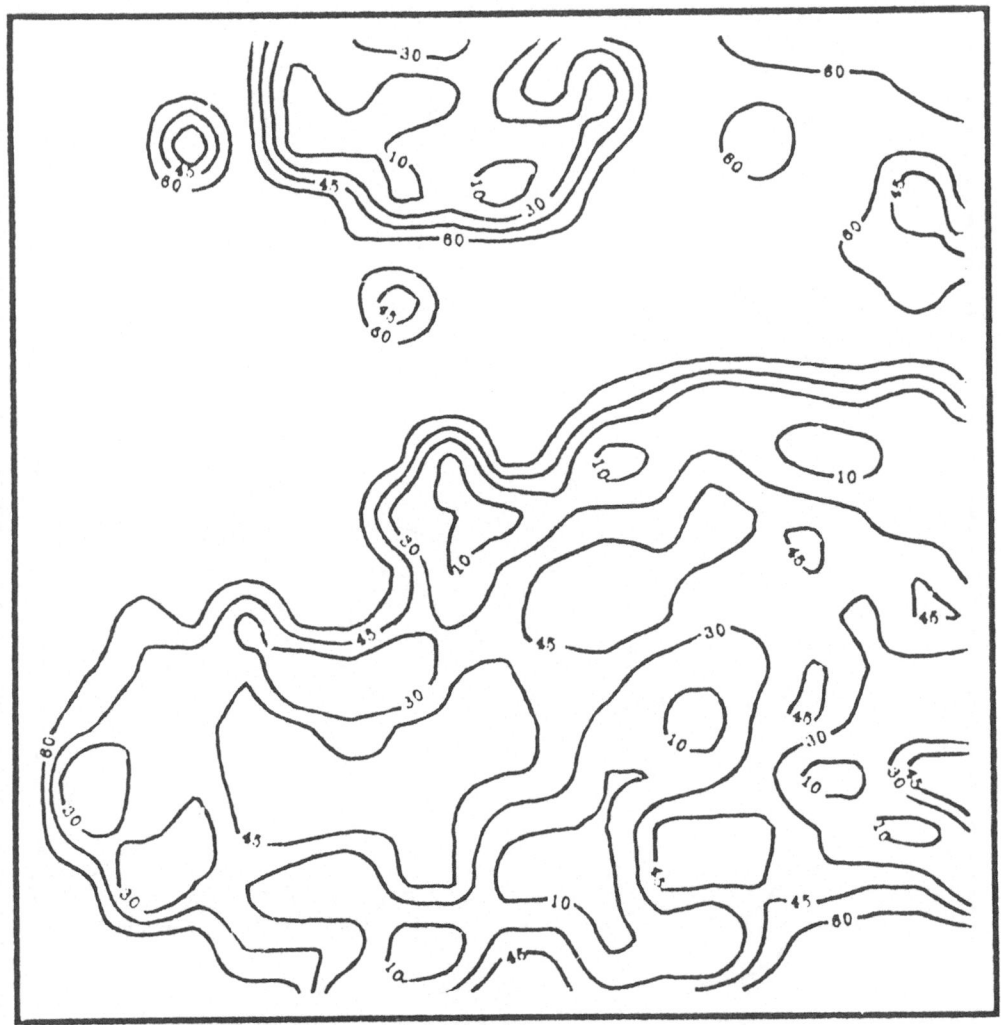

Fig. 16. SO_2-total deposition variances in %; parameter: wind velocity.

– The distributions of partial variances caused by changes in the wind velocity show the known dependence again. Between maximal sensitivities in the neighborhood and at larger distances from the main emissions we can observe a region of minimal parameter influence.

Close to the large source areas the SO_4-total deposition is mainly influenced by variations of the wind velocity. Far away from the sources SO_4-total deposition values react to changes in almost every considered parameter. Only the SO_4-emission portion and the SO_4-deposition velocity are of negligible influence. In the area in between the variations of the SO_2–SO_4-conversion rate become especially important for the calculation of the SO_4-total deposition.

Fig. 17. SO_4-total deposition variances in %; parameter: SO_2–SO_4-conversion rate.

3.5. Sulphur total deposition

– The central statement which can be made about the total Sulphur deposition is that it is mainly influenced by the variations of the wind velocity.

The distributions of the partial variances due to the considered parameters, however, indicate some deficiencies of the 'FAST'-method.

After increasing the value of each single input parameter separately we can recognize positive and negative changes in the deposition field. In the area in between there are zero changes. At these points the total variance caused by simultaneous variation of all parameters is zero or very small, and the 'FAST' method yields results with obvious errors.

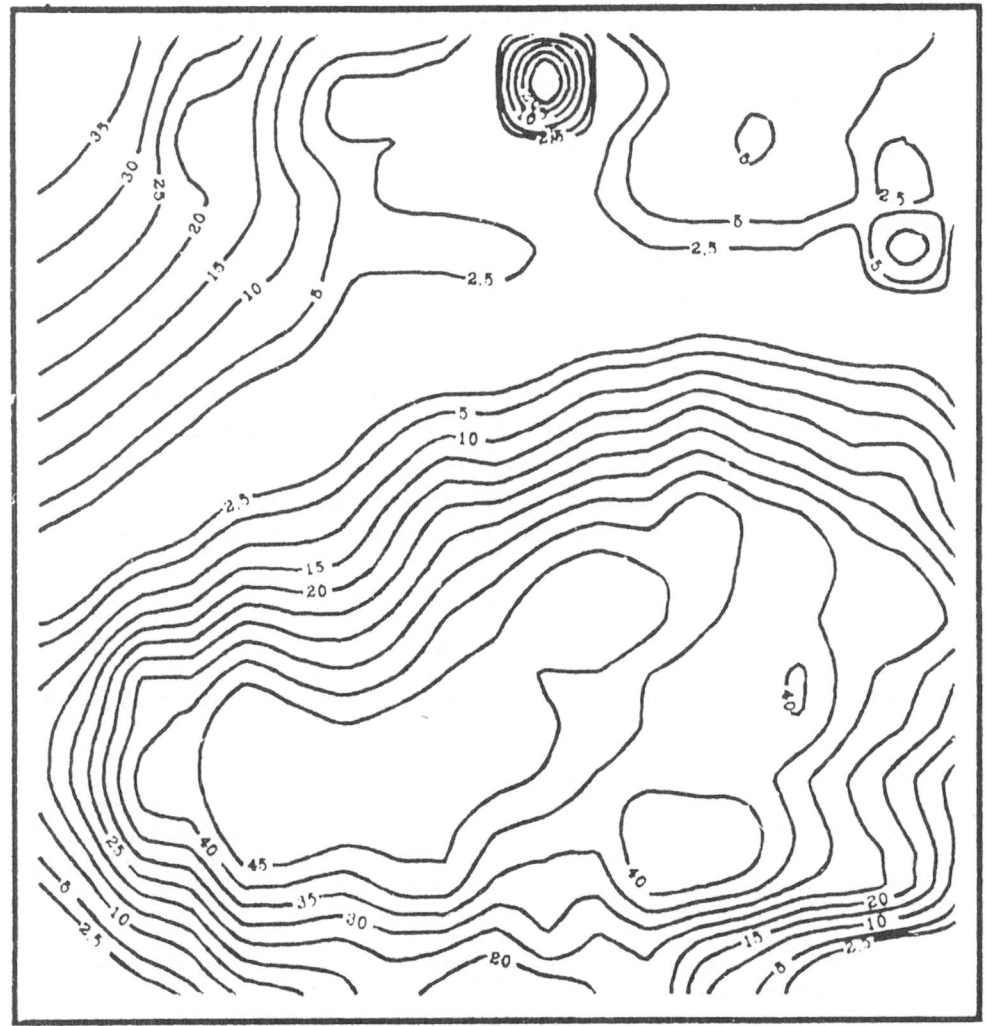

Fig. 18. SO_4-total deposition variances in %; parameter: wind velocity.

Furthermore, Uliaz (1986) refers to the following fact: In the implementation used, the numerator of the partial variances involves the variation of one input parameter, while the total variance in the denominator takes account of the variations of all examined parameters and the coupling between them. So the terms caused by the coupling of parameter variations are not considered and the sum of the partial variances will not equal unity. The contribution of these partial variances may grow with increasing the variation ranges of the input parameters.

In addition, a detailed examination of the application of the 'FAST'-method to our Air Pollution Model had shown a dependence on the accuracy of computation. The sum of partial variances increases with better numerical accuracy. In the present implementation all partial variances were calculated in the double precision mode.

Finally it should be pointed out that all sensitivity statements are made under the assumption of the given limits of parameter variation. If one special input parameter varies in an extended range, it will be obvious that its influence on the output solution increases.

4. Summary

In the present implementation of the 'FAST'-method to the Long Term Interregional Air Pollution Modell we assumed a general range of parameter variations of $\pm 10\%$. On this assumption we recognize that the output variables are widely dominated by variations of the wind velocity. The maximal influence of changes in the SO_4-emission portion and the mixing height close to the sources agrees well with intuition. The variations of the decay rates become important with increasing distance. With this the SO_2–SO_4-conversion rate has a more significant influence on SO_4-output variables than on SO_2-results. The 'FAST'-method, however, requires the combined examination of all partial variances, because the parameters are varied simultaneously. Finally, problems connected with the numerical implementation of the 'FAST'-method are discussed.

References

Cukier, R. I., Fortuin, C. M., Shuler, K. E., Petschek, A. G., and Schaibly, J. H.: 1973, *J. Chem. Phys.* **59**, 3873.
Cukier, R. I., Levine, H. B., and Shuler, K. E.: 1978, *J. Comp. Phys.* **26**, 1.
Cukier, R. I., Schaibly, J. H., and Shuler, K. E.: 1975, *J. Chem. Phys.* **63**, 1140.
Klug, W.: 1982, Physical Transport or the Problem how to model Air Pollution', T. Schneider and L. Grant (eds.), *Air Pollution by Nitrogen Oxides*, Elsevier Scientific Publishing Company, Amsterdam, pp. 243–248.
Klug, W. and Lüpkes, C.: 1985, *Comparison Between Long Term Interregional Air Pollution Models*, Final Report for Umweltbundesamt, Institut für Meteorologie, Technische Hochschule Darmstadt.
McRae, G. J., Tilden, J. W., and Seinfeld, J. H.: 1982, *Chem. Engineering* **6**, 15.
Schaibly, J. H. and Shuler, K. E.: 1973, *J. Chem. Phys.* **59**, 3879.
Uliaz, M.: 1986a, *Fast-Fourier Amplitude Sensitivity Test – Short Manual, Institute of Environmental Engineering*, Technical University of Warsaw.
Uliaz, M.: 1986b, *Computational Implementation of the FAST-Method for Sensitivity Uncertainty Analysis of Air Pollution Long Range Transport Model*, IIASA, Memorandum.

ON COUPLING AIR POLLUTION TRANSPORT MODELS OF DIFFERENT SCALES

ULRICH DAMRATH and RALPH LEHMANN*

Meteorological Service of the GDR, Albert-Einstein-Strasse 42–44, Potsdam 1561, G.D.R.

(Received November 17, 1987; revised July 28, 1988)

Abstract. The effect of coupling a coarse-grid and a fine-grid air pollution transport model (by one-way interaction) is investigated. For this, the application of 'coarse' boundary values (provided by a coarse-grid model) in the fine-grid model is discussed theoretically and demonstrated by test calculations. It turns out that one should always prefer applying coarse information about external (with respect to the fine-grid model) sources to ignoring them.

1. Introduction

In recent years numerous air pollution transport models of various scales have been developed. In the present paper effects of coupling two such models of different scales are investigated. Here we consider a one-way interaction, i.e., a fine-grid model utilizes boundary values provided by a coarse-grid model covering a larger area. Examples for this might be the Operative Air Pollution Transport Model of the G.D.R. (Discher and Damrath, 1986) and the EMEP/MSC–W (or MSC–E) Model covering Europe (Eliassen and Saltbones, 1983; Galperin, 1987). According to Eliassen and Saltbones (1983), about one third of the SO_2 deposited in the G.D.R. comes from foreign sources. That is why it would be desirable to take into account SO_2 fluxes directed into the integration area of the G.D.R. model. However, the corresponding data provided by an interregional model like that of the MSC–W (or MSC–E) have a spatial resolution much coarser than that of the regional G.D.R. model. Thus, the question arises whether the one-way interaction of both the models leads to an improvement of model outputs or not.

For clarity throughout the rest of the paper the fine-grid and coarse-grid model are called 'regional model' and 'interregional model', respectively, having in mind the examples mentioned above.

2. Brief Theoretical Discussion

We suppose that a regional model with horizontal and vertical grid widths Δx, Δy, and Δz shall use boundary values provided by an interregional model with grid widths $\overline{\Delta x}$, $\overline{\Delta y}$, and $\overline{\Delta z}$ $(\overline{\Delta x} = k \cdot \Delta x, \overline{\Delta y} = l \cdot \Delta y, \overline{\Delta z} = m \cdot \Delta z; k > 1, l > 1, m > 1)$. Boundary values are required (by upstream advection methods) at inflow boundaries. If the advection is decomposed into two advection processes (like in 'time splitting' methods) with wind

---

* Author for all correspondence.

vectors parallel and normal to the boundary, respectively, then only the latter one needs boundary values at the corresponding boundary. Let us therefore consider a wind vector perpendicular to the boundary (cf. Figure 1a).

We assume that each grid point represents a box around itself (i.e., a rectangle in two-dimensional projection; cf. Figures 1a, 1b). An upstream advection method, applied to the regional model, would 'expect' boundary values at the imaginary grid

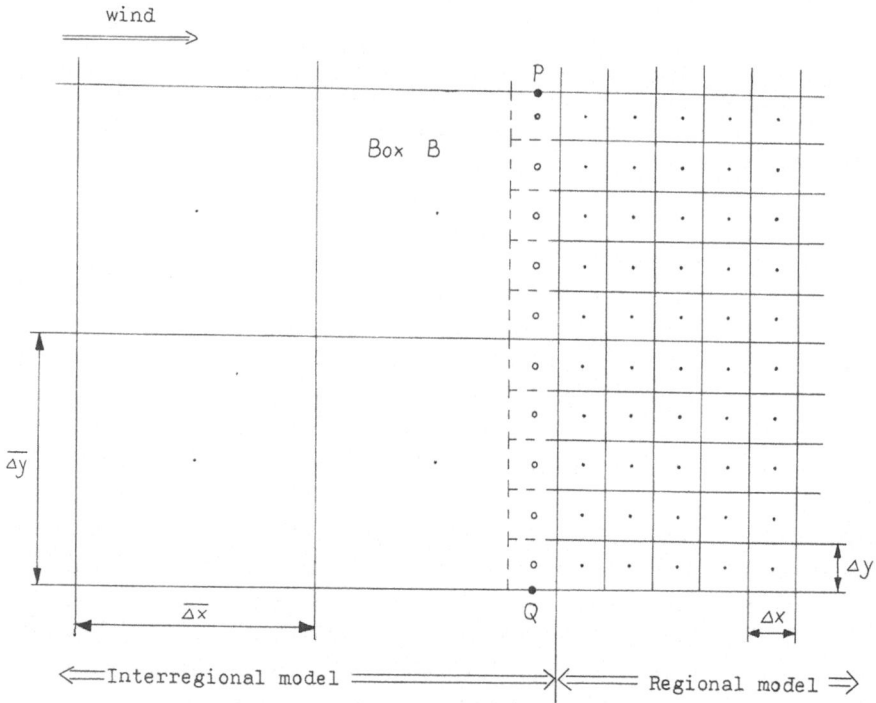

Fig. 1a.   Horizontal structure of the coupled models.

points marked by 'o' in Figures 1a, 1b. We now consider $N = l \cdot m$ such points situated in one $\overline{\Delta x} \cdot \overline{\Delta y} \cdot \overline{\Delta z}$ – box B of the interregional model. 'Ideal' boundary values at these points would be supplied by an interregional model having the same spatial resolution as the regional one; we denote them by $c_1, \ldots, c_N$ (cf. Figure 1b). Because of the coarser spatial resolution, the real interregional model computes an averaged concentration

$$\overline{c} := \frac{1}{N} \sum_{i=1}^{N} c_i \tag{1}$$

for the box B, which corresponds to boundary values

$$\hat{c}_1 = \ldots = \hat{c}_N = \overline{c} \tag{2}$$

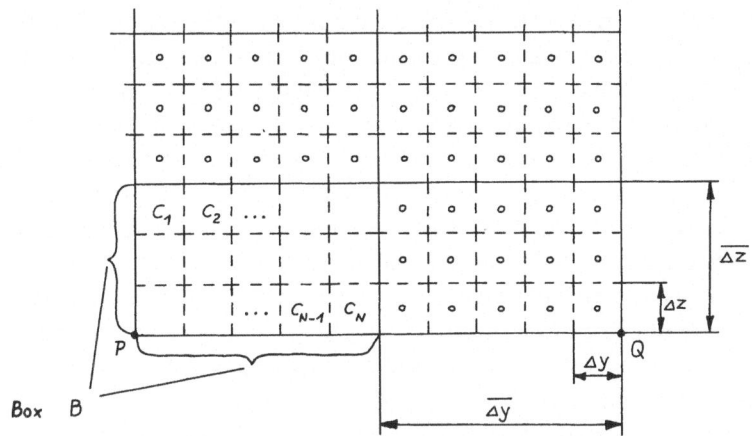

Fig. 1b.   Vertical section along line PQ in Figure 1a.

for the regional model. (Here we have neglected the effects of a possibly different treatment of meteorological and other physical processes in both the models. Moreover, considering in Equation (1) only the effect of averaging in the y-z-plane, we have implicitly assumed that the computed concentrations c_i and $\bar{c}$ are independent of the corresponding grid widths Δx and $\overline{\Delta x}$, i.e., the grid spacing parallel to the wind vector. The latter fact can be easily proved under the assumption of steady state, cf. Appendix A.)

Now we are going to demonstrate that the incorporation of 'coarse' boundary information from the interregional model should be preferred to ignoring the fluxes into the regional model.

The mean square deviation of 'zero' boundary values (ignoring the fluxes into the regional model) from the 'ideal' boundary values is

$$e_0 = \frac{1}{N} \sum_{i=1}^{N} (0 - c_i)^2 = \frac{1}{N} \sum_{i=1}^{N} c_i^2 . \tag{3}$$

The mean square deviation of the boundary values $\hat{c}_i = \bar{c}$ from the 'ideal' ones amounts to

$$\hat{e} = \frac{1}{N} \sum_{i=1}^{N} (\bar{c} - c_i)^2 \tag{4}$$

$$= \frac{1}{N} \sum_{i=1}^{N} c_i^2 - \frac{1}{N} \cdot N \cdot \bar{c}^2$$

$$= e_0 - \bar{c}^2 . \tag{5}$$

It is $\hat{e} < e_0$, i.e., the incorporation of information from the interregional model always results in a reduction of the mean square deviation of the boundary values from the 'ideal' ones (in comparison with ignoring the fluxes into the regional model). Moreover, it follows immediately from Equation (4) that small deviations $\hat{e}$ correspond to a small variance of the concentrations c_i; such concentration profiles are produced especially by sources which are situated far from the boundary (because of diffusion) or by multiple sources (overlaying several profiles results in a smoothed one). The latter effect also occurs if concentration profiles corresponding to a single source, but to various meteorological conditions, are superimposed for the purpose of climatological investigations.

Obviously, the deviations e_0 and $\hat{e}$ of the boundary values can serve only as a clue for estimating the corresponding deviations of the ground level concentrations. That is why numerical tests were carried out for investigating the influence of the model coupling on the accuracy of the computed ground level concentrations. The results are reported in the following chapter.

Note: It has to be stressed that the application of the coarse boundary values in the form described by Equation (2) has only been considered for theoretical purposes. In fact, the conservation of mass at the boundary between the two models does not require the conservation of concentrations as described by Equation (2),

$$\sum_{i=1}^{N} \hat{c}_i = N \cdot \overline{c}, \tag{6}$$

but the conservation of the mass flowing in per time, i.e. the conservation of fluxes,

$$\sum_{i=1}^{N} \hat{c}_i \cdot u_i \cdot A_i = N \cdot \overline{c} \cdot \overline{u} \cdot \overline{A}, \tag{7}$$

where $\hat{c}_i$ is the boundary values for the fine-grid model; u_i, wind speed at the grid point corresponding to $\hat{c}_i$; $A_i = \Delta y \cdot \Delta z_i$, cross-section of the grid box corresponding to $\hat{c}_i$; $\overline{c}$, boundary concentration in the coarse-grid model; $\overline{u}, \overline{A}$, corresponding wind speed and grid-box cross-section in the coarse-grid model.

This is important especially for a strongly spatially varying wind field (including vertical wind shear) and in the case of incompatible wind velocities u_i and $\overline{u}$ resulting from a different treatment of meteorological data in the two models.

3. Numerical Tests

The numerical tests were performed under realistic assumptions on the meteorological conditions over the territory of the G.D.R. during winter. The structure of emissions employed in the tests was adopted from Lehmhaus et al. (1986); particularly, we assumed only one (great) source per $\overline{\Delta x} \cdot \overline{\Delta y} \cdot \overline{\Delta z}$ – grid element, which produces 'unsmoothed' concentration profiles. The models applied can be briefly characterized as follows:

Regional model:
Area covered: $450 \times 600 \text{ km}^2$
Horizontal grid width: 10 km

Vertical structure: 8 levels, upper level at 900 m (which corresponds to the mean mixing height over the G.D.R. in winter, cf. EMP/MSC-W Report 1/83).

Vertical diffusion: based on diffusion coefficients computed from mean vertical temperature profiles for winter days (in dependence on the wind direction, which is correlated with the origin and thus with the thermal stability of the air).

Chemical transformation, dry deposition: computed with values corresponding to SO_2:

$v_d = 0.8$ cm s^{-1},

$K_{chem} = 2.8 \times 10^{-6}$ s^{-1}.

Interregional model:

Area covered: 750×900 km^2

Horizontal grid width: 150 km

Vertical structure: 1 layer, model height: 900 m.

Chemical transformation, dry deposition: as in the regional model.

The following model configurations have been tested:

(A) interregional model with coarse grid (as described above);

(B) no sources outside the regional model (i.e., 'zero' boundary values); and

(C) interregional model substituted by a fine-grid model, equivalent to the regional model (i.e., 'ideal' boundary conditions for the regional model).

The following situations have been simulated:

(a) single case: wind direction, vertical temperature profile etc. were given; and

(b) climatological mean: several concentration profiles obtained in (a) were superimposed according to a frequency distribution of the wind direction.

In order to analyze the test results, we computed the root mean square error (RMSE) (considering model configuration (C) as the 'correct one') of:

(α) the ground level concentrations at single grid points of the regional model; and

(β) the ground level concentrations averaged over 15×15 grid points of the regional model.

TABLE I

Relative mean square errors (expressed as per cent) of the ground level concentration for different model configurations (detailed explanation in Chapter 3).

	Error at single points		Mean error for 15×15 points	
	A (coarse boundary values)	B (zero influx)	A (coarse boundary values)	B (zero influx)
a (single case)	39.	45.	17.	43.
b (climatological)	5.	35.	4.	37.

The results obtained are concentrated in Table I (the values have been normalized such that the mean ground level concentration in the regional model in configuration (C) is 100).

4. Conclusions

It can be seen from Table I that one should always prefer applying coarse information about external sources (A) to ignoring them (B). However, the advantage gained is significantly greater for climatological modelling (b) than for the simulation of single cases (a). The circumstance that for configuration (B) averaging procedures ($\alpha \to \beta$, $a \to b$) yield only a slight improvement of the results is due to the fact that the total mass contained in the model is incorrect in the case of neglected inflow. (This is also reflected by Equation (5): $e_0 = \hat{e} + \bar{c}^2$, where $\bar{c}$ corresponds to the mass flowing in.)

As the dry deposition can be calculated by multiplying the ground level concentration by the deposition velocity (which has been assumed to be constant in these tests), results identical to those in Table I can be obtained for the dry deposition.

If a spatially constant wash-out coefficient is applied, the wet deposition at some grid point (at ground level) is proportional to the vertically integrated concentration ($\int_0^H c(z)\,dz$) above that grid point. That is why one might expect that, for the accuracy of the wet deposition computed, errors in the vertical distribution of pollutants (occurring from the application of coarse boundary values) are not important and, consequently, the errors corresponding to those in Table I are smaller. However, this hypothesis was not confirmed by the numerical tests, which yielded results similar to that in Table I. Probably, this is due to the fact that the pollutants reaching the boundary of the regional model are already relatively well-mixed in the vertical direction. Moreover, the remaining errors in the vertical distribution of pollutants are 'translated' into errors of the horizontal distribution by the effect of vertical wind shear in the regional model.

When assessing the numbers in Table I, one should keep in mind that they have been obtained for a country with a deposition from foreign sources amounting to about one third of the total deposition (for corresponding deposition ratios for other countries see Eliassen and Saltbones, 1983) and that these figures represent only the effect of the different spatial resolution of the regional and the interregional model, excluding the effects of a different treatment of meteorological data.

Appendix A

In order to demonstrate that steady state concentrations computed by an upstream advection scheme do not depend on the grid spacing parallel to the wind vector, we consider a three-dimensional model (with grid widths Δx, Δy, Δz) which contains only one source, starting to operate at the time $t = 0$ with intensity $Q\,[\text{g s}^{-1}]$. The wind of velocity $u\,[\text{m s}^{-1}]$ is assumed to be parallel to the x-axis. We apply the following notations:

c_k = concentration in the $\Delta x \cdot \Delta y \cdot \Delta z$-grid element around the source at the time $t = k \cdot \Delta t$,

$c_\infty = \lim\limits_{k \to \infty} c_k$ ('steady state'),

$\gamma = u \cdot \dfrac{\Delta t}{\Delta x}$ (Courant number),

$\chi = Q \dfrac{\Delta t}{\Delta x \cdot \Delta y \cdot \Delta z}$.

The explicit upstream yields

$c_0 = 0$

$c_{k+1} = \chi + (1 - \gamma) \cdot c_k, \quad k = 0, 1, 2, \ldots,$

and, therefore,

$$c_\infty = \chi [1 + (1 - \gamma) + (1 - \gamma)^2 + (1 - \gamma)^3 + \ldots]$$

$$= \chi \frac{1}{1 - (1 - \gamma)}$$

$$= \frac{\chi}{\gamma}$$

$$= \frac{Q}{u \cdot \Delta y \cdot \Delta z}, \quad \text{which is independent of } \Delta x.$$

References

Discher, H. J. and Damrath, U.: 1986, *Die Einordnung der Modellierung von gasförmigen Luftverunreinigungen in ein Umweltüberwachungssystem*, Abhandlungen des Meteorologischen Dienstes der DDR, Nr. 136, 63–72.

Eliassen, A. and Saltbones, J.: 1983, *Atmospheric Environment* **17**, 1457.

EMEP: 1983, EMEP/MSC-W Report 1/83, The Norwegian Meteorological Institute, Oslo.

Galperin, M. V.: 1987, *Trajectory Model of Long-Range Air Pollutant Transport for the Calculation of Deposition and Concentration*, Proceedings of the WMO Conference on Air Pollution Modelling and its Application (Leningrad, 1986), Technical Document WMO/TD No. 187, pp. 327–334.

Lehmhaus, J., Saltbones, J., and Eliassen, A.: 1986, *A Modified Sulphur Budget for Europe for 1980*, EMEP/MSC-W Report 1/86, The Norwegian Meteorological Institute, Oslo.

CLIMATOLOGICAL VARIABILITY OF SULFUR DEPOSITIONS IN EUROPE

BRAND L. NIEMANN

Acid Rain Staff (OAR-445), Office of Air & Radiation, U.S. Environmental Protection Agency, Washington, D.C. 20460, U.S.A.

(Received November 17, 1987; revised June 8, 1988)

Abstract. The climatological variability in historical and projected S deposition levels for Europe have been simulated using a simple source-receptor model that runs on a personal computer (RCDM) using an extended period of wind and precipitation data. The variability in historical temperature and precipitation data has been analyzed to assess the representativeness of the limited meteorological period used in the EMEP model (1978–1982). A match-up between 40 selected EMEP monitoring sites and the closest climatological station showed the 5-yr average for the EMEP period (1978–1982) and the 35-yr precipitation amounts in generally good agreement for the majority of sites. Comparisons between the RCDM model simulations using the IIASA base 1980 SO$_2$ emissions and the 1978–1982 average precipitation amounts showed the model predictions were generally within a factor of two of the EMEP concentrations and depositions at 40 selected sites. The sensitivity of model evaluation results to 'free parameter' tuning and the appropriateness of the resulting 'free parameters' requires more analysis. The total S depositions at the IIASA receptors predicted by the RCDM model under base year 1980 emissions showed very small differences between the predicted total S depositions for the 1978–1982 EMEP period and the 1951–1985 normal period. The long-period variability in annual total S depositions simulated by the RCDM with constant emissions showed the largest fluctuations in the mid-1970s and showed that the means and C.V.s were not significantly different between the time periods of interest. It is recommended that additional source areas for the Soviet Union be added to the model and the sensitivity to country emission and area centroid locations be explored.

1. Introduction

Forecasting future acid deposition levels on a regional scale is difficult because of the various aspects that need to be considered. First, the future emissions of acid deposition precursors must be forecast for existing and new sources, or specified as a fraction of current of emissions. Second, the meteorological factors that influence regional acid deposition patterns must be specified for the future period of interest. Finally, a regional source-receptor relationship between emissions and depositions must exist for current conditions and must be applicable to future conditions. Considerable progress has been made on these three aspects in both Europe (Alcamo and Bartnicki, 1986) and North America (NAPAP, 1987). Improvements in emission inventories and models that forecast future emissions under various economic growth, energy demand, and environmental control scenarios have been made. The specification of future meteorological factors has heretofore been treated superficially due to insufficient knowledge about meteorological variability and constraints on running models for more than a few years of recent meteorological data. However, the observed year-to-year variability in wet depositions and increased awareness of the abnormalities in precipitation patterns (i.e., drought and floods) has raised concerns about the representativeness of model results based on a very limited period when applied to the future. This concern has prompted analyses of the representativeness of the limited periods of meteorological data that have been used in models relative to conventional climatological normals (usually 30-yr) and efforts to run models for longer periods. The cost and time constraints to prepare the input data bases and run most regional models on main-frame computers for extended periods has in turn prompted the author to develop a simple regional model to run on a personal computer.

The purpose of this paper is to simulate future S deposition levels for Europe under various alternative emission reduction plans using a simple model on a personal computer (PC). The purpose of this paper is to also analyze the variability in basic meteorological factors that influence acid deposition patterns and assess the representativeness of limited meteorological periods that have been used heretofore (i.e., 1978–1982). A description of the PC-based data analyses and source-receptor model are presented in the next section. The third section contains meteorological variability and representativeness results using temperature and precipitation data bases and the model evalution results using the European Monitoring and Evaluation Programme data base (EMEP, 1983). The forth section contains the model application results using base year and emission reduction scenarios and a 30-yr wind and precipitation data base to calculate the year-to-year variability in S depositions and concentrations at selected receptors. The conclusions and recommendations are presented in the last section.

2. PC-Based Data Analysis and Model System

2.1. DATA BASE ANALYSES

An IBM PC/AT computer and the LOTUS 1–2–3 spreadsheet software program were used for both data analyses and the source-receptor model. The spreadsheet program was used to analyze the data bases and prepare the model inputs at the same time and to analyze the model outputs. The large main-frame data bases for historical upper air wind and precipitation amounts were extracted for the stations of interest, processed for monthly, seasonal, and annual values, reformatted, and 'downloaded' to PC/LOTUS format. The smaller data bases for emissions, monitoring data, etc., were key entered directly into the LOTUS spreadsheets.

The source region inputs to the source-receptor model are summarized in Table I. An upper air station was assigned to each source region and the annual resultant winds from the U.S. National Climatic Data Center upper air digital files (NCDC, 1985) were computed for each station-year with at least 9 mo of monthly resultant winds at the 850 mb level.

The IIASA (Hordijk, 1986, 1987) receptor data for the model are summarized in Table II. The closest station with a long period of record in the World Monthly Surface Station Climatology (Spangler and Jenne, 1985) to each of the 12 receptors was selected and the data was analyzed for the mean and coefficient of variation (C.V.) in precipitation amounts. A C.V. of greater than about 0.15 is generally considered significant and

TABLE I

Source region inputs to the Regional Climatological Deposition Model

Region	EMEP Source region	Country abbrev.	Country name	Area[a] ($\times 10^5$ km^2)	1980[c] SO$_2$ (TgS)	Upper air station
Central	1	A	Austria	0.84	0.159	Wein
	2	CS	Czechoslovakia	1.28	1.832	Libius
	3	D	FR Germany	2.49	1.602	Stuttgart
	4	DDR	DR Germany	1.08	2.415	Lindenberg
	5	L	Luxemborg	0.03	0.020	Nancy
	6	CH	Switzerland	0.41	0.067	Payerne
North	7	DK	Denmark	0.43	0.226	Kobenhavn
	8	SF	Finland	3.37	0.294	Jokionen
	9	IS	Iceland	1.03	0.006	Keflavik
	10	N	Norway	3.24	0.072	Oslo
	11	S	Sweden	4.50	0.243	Kobenhavn
West	12	B	Belgium	0.31	0.432	Uccle
	13	F	France	5.47	1.657	Trappes
	14	IRL	Ireland	0.70	0.119	Valentia
	15	NL	Netherlands	0.41	0.234	De Bilt
	16	UK	United Kingdom	2.44	2.342	Aughton
South	17	AL	Albanië	0.29	0.039	Zagreb
	18	BG	Bulgaria	1.11	0.508	Bucuresti
	19	GR	Greece	1.32	0.345	Athens
	20	I	Italy	3.01	1.898	Milano
	21	P	Portugal	0.92	0.130	Madrid
	22	E	Spain	5.05	1.879	Madrid
	23	TR	Turkey	7.81	0.497	Ankara
	24	YU	Yugoslavia	2.56	0.837	Zagreb
East	25	H	Hungry	0.93	0.813	Budapest
	26	PL	Poland	3.13	1.741	Legionowo
	27	R	Romania	2.38	0.757	Bucuresti
	28	SU	Soviet Union	50.44[b]	8.588	Minsk
Total				106.98	29.76	

[a] *National Geographic Atlas* (1985), Washington D.C., U.S.A.
[b] European portion.
[c] Hordijk (1986).

was found at 11 of the 12 sites. This means there could be an inherent variability in annual depositions of about ± 0.15 just due to variations in precipitation amounts themselves with constant emissions and transport conditions. In addition, the annual temperatures and precipitation amounts at 16 long-period sites in Europe during 1874–1985 and at 88 sites during 1951–1985 were analyzed for the mean and C.V. and annual normality factors at individual sites and averaged over the region.

The ambient SO$_2$ and SO$_4^{2-}$ concentrations and wet S depositions and precipitation mounts at 40 of the 82 EMEP sites with the most complete concurrent air quality and

TABLE II

IIASA receptor data for the regional climatological deposition model

IIASA receptor name and countries [a]	Lat.	Long. (degrees)	Surface WWD station (period of record) [b]	Lat.	Long.	Norm PRCP (mm)	C.V. [c] none
A Erzgebirge, GDR and CS	51	13E	Cheb, CS (1953–1985)	50 05	12 24	553	0.19
B Katowice, PL	50	19E	Ostrava, CS (1951–1985)	49 47	18 16	715	0.15
C Donetz, SU	48	38E	Rostow, SU (1951–1985)	47 15	39 49	547	0.20
D Rhineland, FRG	51	7E	Essen, FRG (1951–1985)	51 25	06 57	907	0.18
E Fichtel Gebrige, FRG	50	12E	Nuernberg, FRG (1955–1985)	49 30	11 05	634	0.16
F Bilo Gora, YU	46	17E	Zagreb/Gric, YU (1951–1985)	45 49	15 59	893	0.16
G Moscow, SU	56	39E	Moscow, SU (1951–1985)	55 45	37 34	633	0.18
H Lombardy, I	46	09E	Milano, I (1951–1985)	45 28	09 17	965	0.22
I West Yorkshire, UK	53	02W	Waddington, UK (1951–1985)	53 10	0 31W	604	0.17
J Black Forest, FRG	49	08E	Stuttgart, FRG (1952–1970)	48 41	09 12	729	0.13
K Borzsony Hills, H	48	20E	Budapest, H (1951–1985)	47 31	19 01	590	0.19
L Belgrade, YU	45	21E	Szeged, H (1951–1985)	46 15	20 09	505	0.17

[a] Hordijk (1987).
[b] U.S. National Climatic Data Center/National Center for Atmospheric Research World Weather Surface Data File (1985).
[c] Coefficient of variability in annual precipitation amounts (standard deviation divided by the period of record mean).

precipitation chemistry data were extracted from the published data. The 40 sites were selected to include at least one site in each of the 22 countries with EMEP sites for spatial representativeness. It was assumed that the published EMEP data had been screened for completeness. The EMEP data base was entered in a spreadsheet in a format for use in both model evaluation and for calculation of the wet scavenging ratios. The region average calculated scavenging ratio was used as an input to the model in the calculation of wet depositions.

2.2. SOURCE-RECEPTOR MODEL

The Regional Climatological Deposition Model (RCDM), developed under support from the U.S. Environmental Protection Agency and applied by others using mainframe computers, was adapted to a standard input data set from an international model intercomparison and to the **IBM PC LOTUS** 1–2–3 spreadsheet system (Niemann, 1985, 1986). The RCDM is an adaptation of an model published earlier (Fay and Rosenzweig, 1980). The input data files (i.e., emissions, resultant winds, precipitation amounts, etc.) were preprocessed (and analyzed) in separate 1–2–3 spreadsheets and given LOTUS 'range-names' so they could be readily transferred to the main spreadsheet program by the LOTUS file-combine-copy feature. The operation of the main spreadsheet program was automated by a **LOTUS MACRO** that contains the stored sequence of key strokes needed to execute the complete model from start to finish unattended (LOTUS, 1985). The main spreadsheet program requires about 1000 KB of RAM (random access memory) which was made possible by the installation of an expanded memory board in the PC.

Two versions of the PC LOTUS RCDM were developed and used for the EMEP region: the first for the 40 selected EMEP monitoring sites, and the second for 40 receptors consiting of the 28 EMEP countries plus the 12 IIASA receptors (recall Tables I and II). The former was used for model evaluation and sensitivity analyses to the model 'free parameters' (eddy diffusivity and residence time scales) while the latter was used for model applications.

3. Results

3.1. METEOROLOGICAL VARIABILITY

Surface weather observations have been made in Europe for a long period of time at some sites. The World Monthly Surface Station Climatology contains a number of stations with very complete records back to the mid-1800s. The average normality factors for the 16 long-period sites in Europe for 1874–1985 in Figure 1 showed several extended periods of abnormal precipitation amounts (i.e., the 1945–1950 drought). For the most recent period corresponding to deposition measurements, more stations are available for all of Europe. The average normality factors for precipitation amounts based on 88 sites in Figure 2 showed years like 1953 and 1966 had the most abnormal precipitation amounts regionally. The abnormality during 1966 was even more pronounced and of longer duration in the central region. It should be noted that the precipitation normality factors for the most recent 15 yr are probably biased low due to the fact that 33 of the 88 stations had incomplete data and the average amounts were substituted in those years. This situation can be rectified when the data from these sites are published.

The annual normality factors for temperature and precipitation were combined in a scatter plot to see how symmetrical the distribution was and to identify years that were clearly different in either temperature and/or precipitation amounts. The plots in Figures 3 and 4 showed a fairly symmetrical distribution about the normal intercepts with value 1.0, but with several years that stand out from the rest as being much warmer and dryer than the other years (i.e., 1921, 1934, 1949, and 1953). Interestingly, none of the years used in the EMEP model (1978–1982) stood out as being especially abnormal from this analysis.

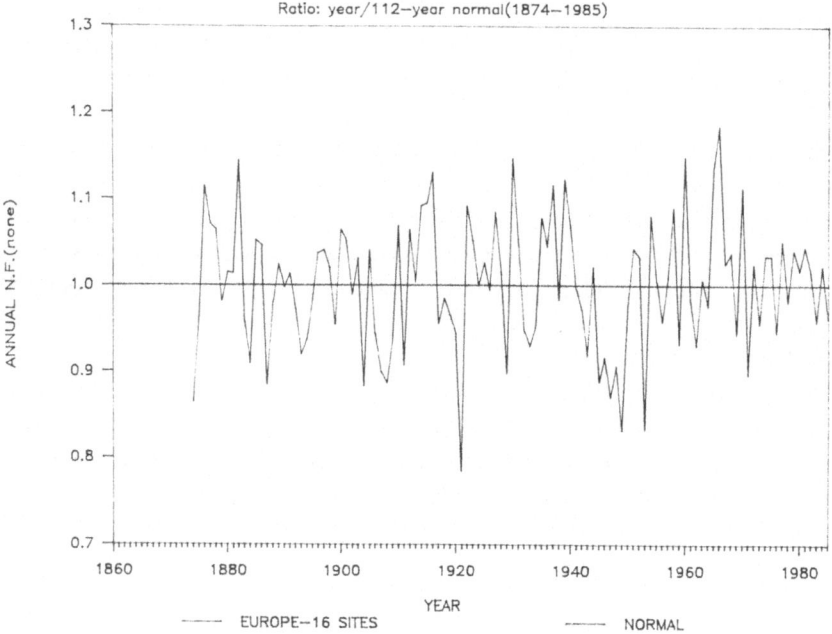

Fig. 1.   Climatological variability in annual normality factors for precipitation at 16 long-period sites in Europe.

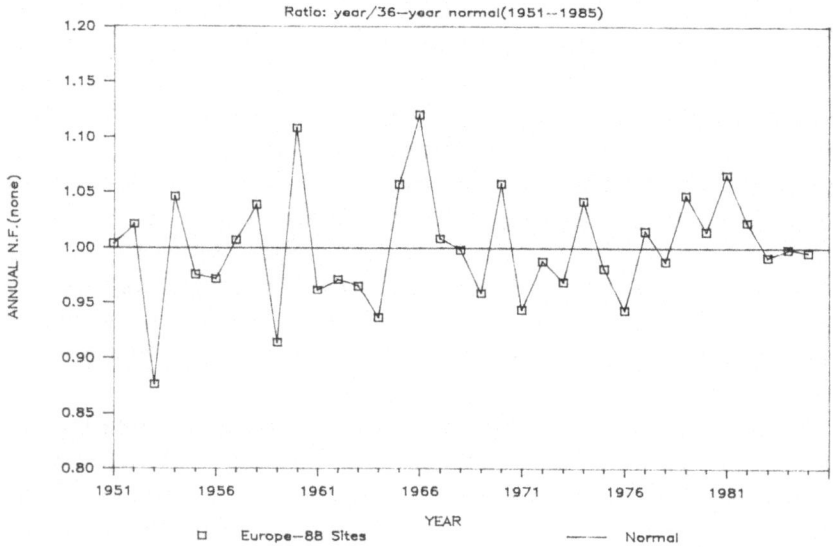

Fig. 2.   Climatological variability in annual normality factors for precipitation amounts at 88 sites in Europe.

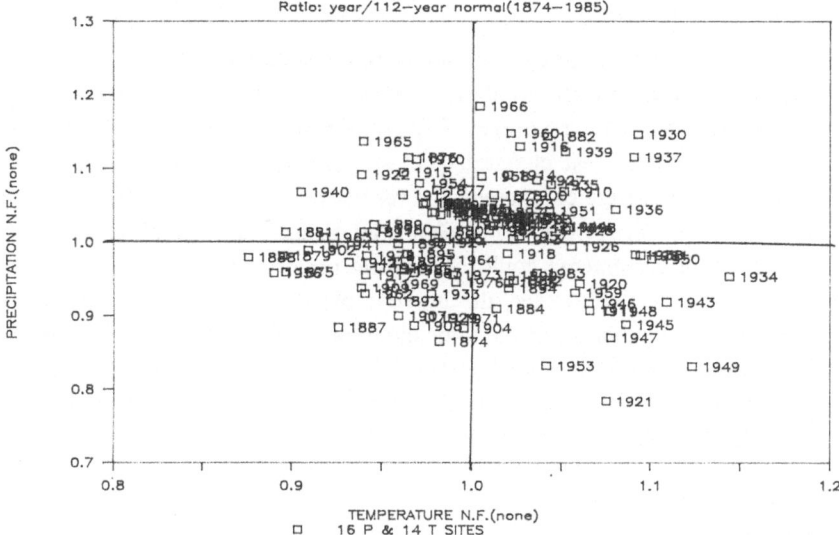

Fig. 3.   Climatological variability in annual normality factors for precipitation amounts versus temperatures at long-period sites in Europe.

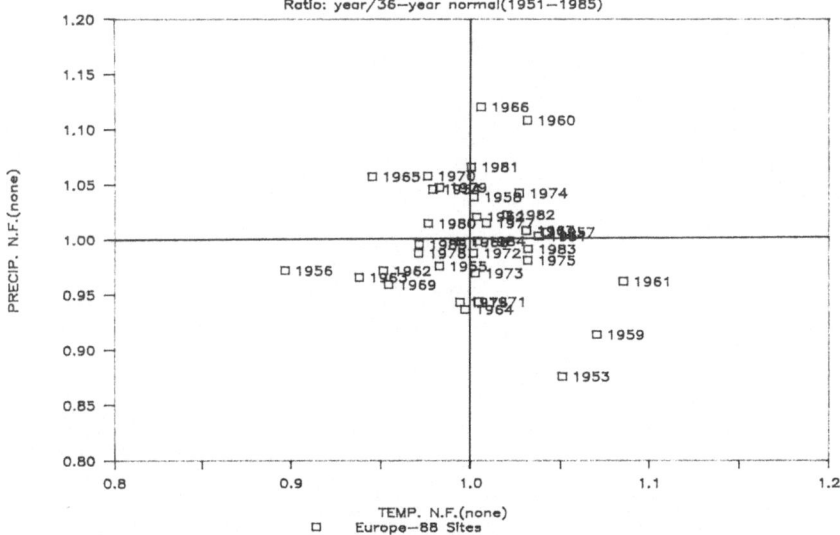

Fig. 4.   Climatological variability in annual normality factors for precipitation amounts versus temperatures at 88 sites in Europe.

3.2. Representativeness

Since the EMEP monitoring data and modeling results are for the period 1978–1982, it is of interest to see how representative the precipitation amounts were during that period compared to the 35-yr normals. Using the match-ups between the 40 selected EMEP sites and the closest climatological station, the 5-yr average and the 35-yr precipitation amounts were compared. The results in Figure 5 showed generally good agreement for the majority of sites, but significant disagreement for about 10 sites. The reasons for the disagreements could be missing data, differences in gage exposure, intervening terrain, etc. However, the intercomparison between the 5-yr and 35-yr average precipitation amounts at the same climatological sites showed excellent agreement for all but a few sites in Figure 6. This result leads one to conclude that the 5-yr EMEP period probably provides deposition results that are representative of the longer period.

3.3. Model evaluation

The RCDM was run with the data inputs as described previously and the model 'free parameters' as follows: eddy diffusivity, 6.0×10^5 m^2 s^{-1}, and primary pollutant, chemical conversion, and secondary pollutant time scales of 8.0×10^4 s, 2.9×10^5 s, and 7.1×10^4 s, respectively, based on previous experience with model evaluations in North America. The SO$_2$ and SO$_4^{2-}$ background concentrations were specified as 0.1 and 0.5 µg S m^{-3}, respectively, the dry deposition velocities as 0.008 and 0.002 m s^{-1}, respectively, and the scavenging ratios as 1.0×10^4 and 7.9×10^5, respectively, based on the European modeling experience (Eliassen and Saltbones, 1983) and data analysis.

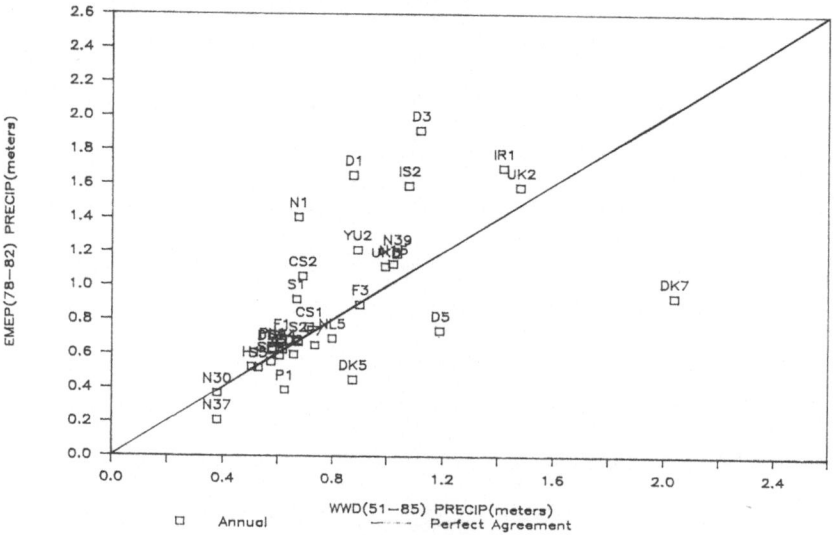

Fig. 5.   Comparison of annual precipitation amounts at 40 EMEP sites for the 5-yr period (1978–1982) and at the closest climatological site for the 35-yr period.

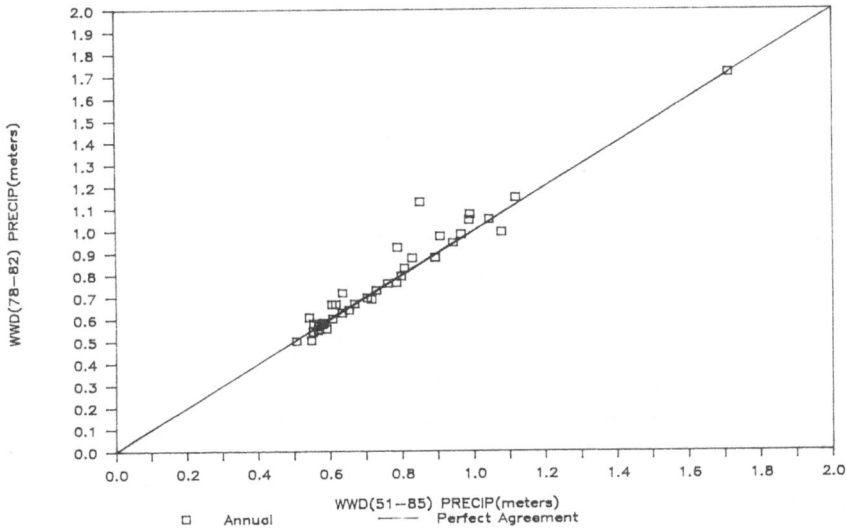

Fig. 6. Comparison of annual precipitation amounts for the 5-yr period (1978–1982) and the 35-yr period (1951–1985) at the climatological sites for the 28 EMEP countries and 12 IIASA receptors.

It should be noted that the background concentrations used were considerably less than those published recently (Szepesi and Fekete, 1987).

The RCDM simulations using the IIASA base 1980 SO_2 emissions (Hordijk, 1987) and the 1978–1982 average precipitation amounts were compared to the EMEP monitoring data. It was found that the RCDM seriously overpredicted the SO_2 concentrations at 9 sites while the predictions and observations were generally within a factor of two at the other sites with SO_2 data. The overprediction is probably due to the close proximity of the emission centroid and the monitoring site. The RCDM predictions for SO_4^{2-} concentrations were generally within a factor of two except for some outliers and a group of sites with SO_4^{2-} concentrations less than 6 $\mu g\ m^{-3}$ for which the model seriously underpredicted. The RCDM predictions for wet depositions were also generally within a factor of two of the observations except for some outliers.

In general one can almost always produce a better fit to a data set by adjusting the model 'free parameters' as was done for the wet S depositions in Figure 7, but the parameter values required to do this seemed out of line with those in the published literature so they were not used. The subject of sensitivity of model evaluation results to 'free parameter' tuning and the appropriateness of the resulting 'free parameters' certainly requires more analysis. In addition, since the scavenging ratios for SO_2 (see Figure 8) and SO_4^{2-} (not shown) were both found to show a dependence on SO_2 concentrations, the use of this empirical dependence should provide a better fit to observations and a simple way of treating 'nonlinearity' (Alcamo et al., 1987).

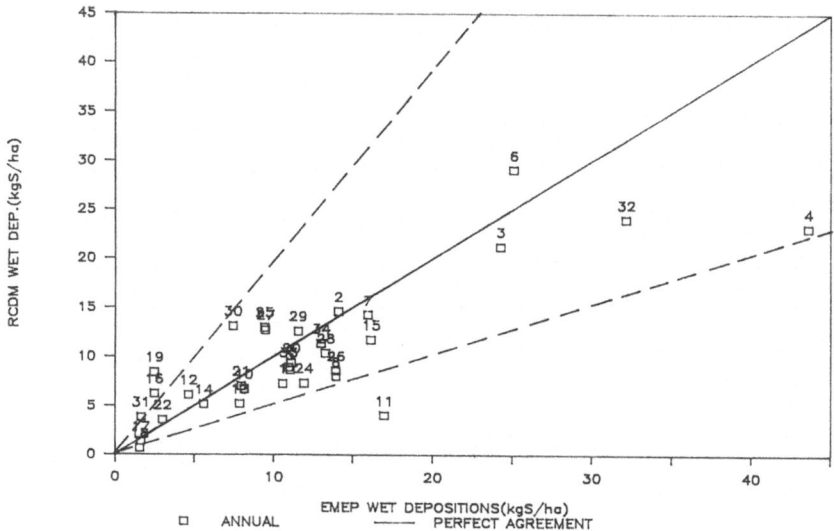

Fig. 7.   Comparison between the EMEP monitored and the 'tuned' RCDM modeled wet sulfur depositions
for the 1978–1982 period.

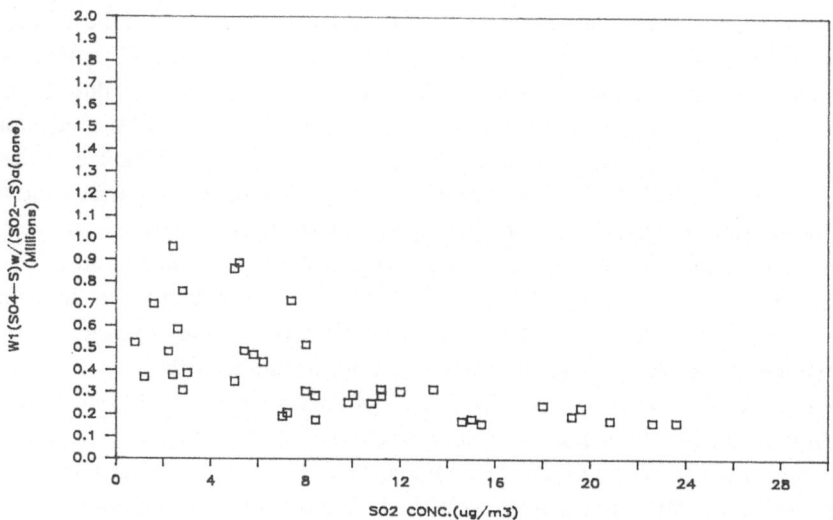

Fig. 8.   Wet scavenging ratio for SO_2 versus SO_2 concentration from the EMEP data for 1978–1982
period.

4. Model Applications

4.1. PREDICTIONS FOR REDUCTION SCENARIOS

The total S depositions at the IIASA receptors simulated by the RCDM model under year 1980 base and year 2000 reductions are shown in Table III. The emissions used in the model simulations are summarized by large regions in Table IV (recall Table I for countries in each region). The base and reduction emissions were developed by IIASA (Hordijk 1986, 1987) and the latter are defined as follows: CRP – current reduction plans; RBI – reductions based on indicators; and TER – targeted emission reductions. It may be noted that the total emissions under the largest reduction scenario correspond to the past level of total emissions in about the late 1960s. Thus it is important to run and evaluate the model against any available monitoring data for that period to check the model applicability to such a large reduction (about 43%) from current conditions.

TABLE III

Total S depositions in $g\,m^{-2}$ at the IIASA receptors from the Regional Climatological Deposition Model

IIASA receptor name and countries [a]	IIASA Base/P7882 [b]	IIASA Base/Pnorm [c]	IIASA CRP/Pnorm	IIASA RBI/Pnorm	IIASA TER/Pnorm
A Erzgebirge, GDR and CS	12.3 (16.8) [a]	12.3	8.6 (11.7)	3.3	7.4
B Katowice, PL	7.3 (12.7)	7.5	6.5 (11.4)	4.0	4.1
C Donetz, SU	0.8* (11.9)	0.9*	0.7* (11.2)	0.6*	0.6*
D Rhineland, FRG	4.2 (9.2)	4.2	2.3 (5.3)	1.8	2.4
E Fichtel Gebrige, FRG	11.9 (8.8)	11.7	7.9 (5.9)	3.5	6.8
F Bilo Gora, YU	9.8 (7.8)	10.1	8.2 (6.8)	5.2	5.4
G Moscow, SU	0.7* (7.3)	0.7*	0.5* (6.8)	0.5*	0.4*
H Lombardy, I	10.4 (6.2)	10.6	7.0 (4.4)	5.3	5.7
I West Yorkshire, UK	9.2 (5.8)	9.1	9.1 (5.6)	5.8	4.6
J Black Forest, FRG	10.1 (5.0)	10.3	5.3 (2.9)	4.1	5.5
K Borzsony Hills, H	7.4 –	7.6	6.3 –	3.9	4.1
L Belgrade, YU	4.9 –	5.0	4.2	2.8	2.9

[a] Hordijk (1986, 1987).
[b] 1978–1982 precipitation.
[c] 1951–1985 precipitation.
* Does not include local source contribution.

TABLE IV

Historic and projected SO$_2$ emissions by European region (TgS)

Region	IIASA base 1980 SO$_2$	IIASA CRP 2000 SO$_2$	IIASA RBI 2000 SO$_2$	IIASA TER 2000 SO$_2$	1950[a] SO$_2$	1972 SO$_2$	1978 SO$_2$
Central	6.095	3.755	1.866	3.292	3.565	6.518	5.590
North	0.841	0.387	0.428	0.793	0.194	1.083	0.850
West	4.793	3.603	2.683	2.602	4.290	5.593	5.065
South	6.133	5.410	3.940	4.532	0.941	2.350	5.225
East	11.899	9.079	7.970	6.748	1.286	3.510	11.350
Total	29.761	22.234	16.887	17.967	10.276	19.054	28.080

[a] Fisher (1983).

The results in Table III showed there were very small differences between the predicted total S depositions for the 1978–1982 EMEP period and the 1951–1985 normal period. However, Table III also showed significant differences (factor of two) between the RCDM and IIASA/EMEP model predictions at some of the twelve receptors. Part of the difference is undoubtedly due to the use of only one source region for the Soviet Union in the RCDM while the EMEP used a gridded inventory. It should be noted that the two IIASA receptors in the Soviet Union show only predicted background total S depositions since the local source contributions of these regions, which are both very large (NILU, 1984), were not included in the RCDM model simulations. The reasons for the differences between the two model results need to be explored further.

4.2. LONG-PERIOD VARIABILITIES

The long-period (1951–1985) annual variability in depositions and concentrations were simulated by the RCDM using the historical wind (1960–1983) and precipitation (1951–1985) data base. In these simulations the SO$_2$ emissions were held constant to isolate the effects of meteorological variability. In future model simulations the actual estimated historical emissions could be used (Table IV). In these simulations the long-period average resultant winds were used for the 1951–1959 and 1984–1985 periods when the wind data were not available. Wind data for the 1984–1986 period are available in published from and could be used to update the wind data file in the future.

The long-period variability in annual wet and total S deposition simulated by the RCDM with constant emissions and other model inputs are shown in Figures 9 and 10 for IIASA receptors selected so as to not overlap appreciably for clarity in the plots. As expected the wet S depositions showed large fluctuations at sites that showed large fluctuations in predicted SO$_4^{2-}$ concentrations and experienced large fluctuations in precipitation amounts (recall the C.V.s in Table II). The total S depositions showed fluctuations during the entire period at most of the selected receptors and showed the largest fluctuations in the mid-1970's.

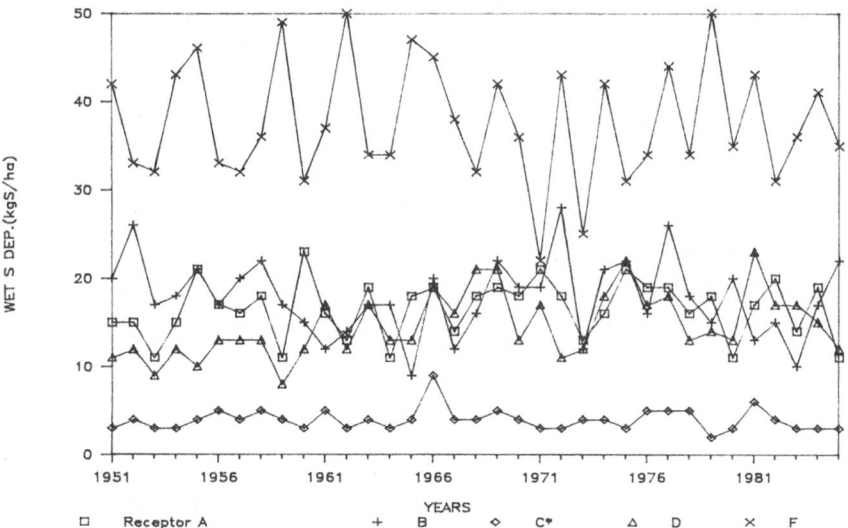

Fig. 9. Variability in annual wet S depositions at selected IIASA receptors simulated by the RCDM model.

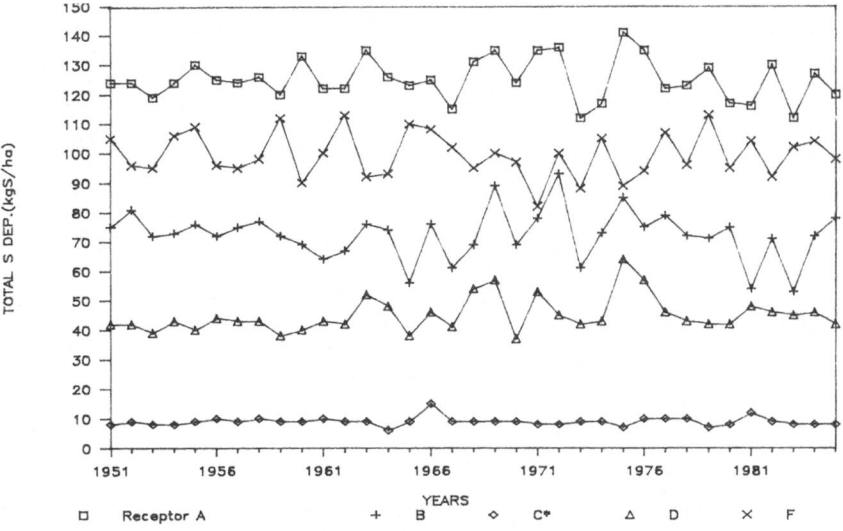

Fig. 10. Variability in annual total S depositions at selected IIASA receptors simulated by the RCDM model.

The variability in annual total S depositions and SO_4^{2-} concentrations at the IIASA receptors predicted by the RCDM model has been summarized in Table V in terms of the mean and C.V. for three time periods. In general the results in Table V showed that the means and C.V.s for total S deposition were not significantly different among the three time periods, but varied more for the SO_4^{2-} concentrations. The 1960–1983 period in Table V was used because it corresponds to the period when both winds and precipitation amounts were varying while the 1951–1985 period includes shorter periods when the winds were constant (1951–1959 and 1984–1985) due to lack of data. It should be noted that the differences between the predictions in Table III using the normal average winds and precipitation amounts and those in Table V using the 35 individual years are small (due to rounding errors) as one would expect if the calculations were done correctly.

TABLE V

Variability in total S depositions and sulfate concentrations at the IIASA receptors from the Regional Climatological Deposition Model as a function of averaging period

IIASA receptor name and countries [a]	Totals 1951–1985 [d]	Totals 1960–1983 [d]	Totals 1978–1982 [d]	SO_4^{2-} 1951–1985	SO_4^{2-} 1960–1983 [e]	SO_4^{2-} 1978–1982
A Erzgebirge, GDR and CS	125 [b]	126	123	10.0	10.6	9.7
	0.06 [c]	0.06	0.05	0.21	0.22	0.30
B Katowice, PL	72	71	69	8.7	8.4	8.2
	0.12	0.14	0.11	0.22	0.26	0.17
C Donetz, SU	9*	9*	9*	2.8*	2.8*	3.1*
	0.17	0.19	0.19	0.19	0.23	0.28
D Rhineland, FRG	45	46	44	5.9	6.4	5.8
	0.13	0.14	0.05	0.27	0.26	0.17
E Fichtel Gebrige, FRG	105	100	103	15.3	13.1	13.2
	0.12	0.13	0.04	0.36	0.41	0.16
F Bilo Gora, YU	99	99	100	15.3	15.1	15.8
	0.08	0.08	0.08	0.11	0.13	0.15
G Moscow, SU	7*	7*	8*	1.8*	1.9*	2.2*
	0.25	0.29	0.27	0.35	0.39	0.25
H Lombardy, I	103	102	101	16.0	15.6	15.3
	0.10	0.10	0.07	0.13	0.16	0.01
I West Yorkshire, UK	92	93	95	5.5	5.7	6.7
	0.04	0.04	0.07	0.24	0.27	0.34
J Black Forest, FRG	100	99	96	12.4	11.8	11.1
	0.07	0.08	0.06	0.18	0.21	0.22
K Borzsony Hills, H	80	82	84	10.8	12.0	13.1
	0.08	0.09	0.10	0.35	0.34	0.37
L Belgrade, YU	49	49	49	11.3	11.1	11.1
	0.07	0.08	0.11	0.15	0.18	0.24

[a] Hordijk (1986, 1987).
[b] Average over period.
[c] Coefficient of variation (standard deviation divided by the average).
[d] kg S ha^{-1}.
[e] µg m^{-3}.
* Does not include local source contribution.

5. Conclusions and Recommendations

Future S deposition levels for Europe under various alternative emission reduction plans have been simulated using a simple source-receptor model on a personal computer (PC) and a climatologically representative period of wind and precipitation data. The variability in basic meteorological factors that influence acid deposition patterns has been analyzed to assess the representativeness of limited meteorological periods that have been used heretofore (i.e., 1978–1982).

The coefficient of variation in annual precipitation amounts at the closest climatological station to the 12 IIASA receptors was found to be greater than about 0.15 which means there could be an inherent variability in annual depositions of about ± 1.15 due to just variations in precipitation amounts themselves with constant emissions and transport conditions. The average normality factors for long-period sites in Europe for 1874–1985 showed several extended periods of abnormal precipitation amounts (i.e., 1945–1950 drought). The abnormality during 1966 was even more pronounced and of longer duration in the central region of Europe. Using a match-up between 40 selected EMEP sites and the closest climatological station, the 5-yr average for the EMEP period (1978–1982) and the 35-yr precipitation amounts showed generally good agreement for the majority of sites.

The comparisons between the RCDM model simulations using the IIASA base 1980 SO_2 emissions and the 1978–1982 average precipitation amounts showed the RCDM simulations for SO_4^{2-} concentrations were generally within a factor of two as were the RCDM predictions for wet S depositions. The subject of sensitivity of model evaluation results to 'free parameter' tuning and the appropriateness of the resulting 'free parameters' requires more analysis.

The total S depositions at the IIASA receptors simulated by the RCDM model under base year 1980 emissions showed very small differences between the predicted total S depositions for the 1978–1982 EMEP period and the 1951–1985 normal period. However, there were significant differences (factor of two) between the RCDM and IIASA/EMEP model predictions at some of the twelve receptors that need to be explored further. The long-period variability in annual total S depositions simulated by the RCDM with constant emissions and other model inputs showed the largest fluctuations in the mid-1970s and showed that the means and C.V.s were not significantly different between the time periods of interest. Additional source areas for the Soviet Union should be added to the model and the sensitivity to country emission and area centroid locations should be explored.

Acknowledgments

The upper air wind and precipitation data bases were processed and downloaded from the main frame computer by Bill Hamilton. The author benefited from helpful discussions with Harald Dovland and Anton Eliassen and suggestions by Joe Alcamo.

Disclamer

Although the author has produced this paper as an employee of the U.S. Environmental Protection Agency, it has not been subjected to Agency peer review and therefore does not necessarily reflect the views of the Agency and no official endorsement should be inferred.

References

Alcamo, J. and Bartnicki, J. (eds.): 1986, *Atmospheric Computations to Assess Acidification in Europe: Work in Progress, RR-86-5,* International Institute for Applied Systems Analysis, Laxenburg, Austria, November, 93 pp.

Alcamo, J., Bartnicki, J., and Schopp, W., 1987: *Effect of Nonlinear Sulfur Removal on Computed Sulfur Source-Receptor Relationships: Some Model Experiments, RR-87-20, Interregional Air Pollutant Transport: The Linearity Question,* International Institute for Applied Systems Analysis, Laxenburg, Austria, November, 83–89.

Eliassen, A. and Saltbones, J.: 1983, *Atmos. Environ.* **17**, 1457.

Fay, J. A. and Rosenzweig, J. J.: 1980, *Atmos. Environ.* **14**, 355.

Fisher, B. E. A.: 1983, *Atmos. Environ.* **17**, 1865.

Hordijk, L.: 1986, *Atmos. Environ.* **20**, 2053.

Hordijk, L.: 1987, 'Acid Rain Abatement Strategies in Europe', in T. Schneider (ed.), *Acidification and its Policy Implications,* Elsevier Science Publishers B.V. Amsterdam, The Netherlands, pp. 295–305.

LOTUS Development Corporation: 1985, *1-2-3 Reference Manual (Release 2).*

National Acid Precipitation Assessment Program: 1987, *Interim Assessment: The Causes and Effects of Acidic Deposition,* Volume I – *Executive Summary,* Office of the Director of Research, 722 Jackson Place, NW, Washington, D.C. 20503.

National Geographic Atlas: 1955, National Geographic Society, Washington, D.C.

National Climatic Data Center Upper Air Digital Files TD-6200 Series: 1985, Asheville, North Carolina, June, 16 pp.

Niemann, B. L.: 1985, *Evaluation of the ISDME Data Set and a Regional Climatological Deposition Model,* Paper 85–5.4, Proceedings of the 78th Annual Meeting of the Air Pollution Control Association, Pittsburgh, Pennsylvania, 23 pp. and Appendix.

Niemann, B. L.: 1986, 'Regional Acid Deposition Calculations with the IBM PC LOTUS 1-2-3 System', *International Quarterly Journal of Environmental Software,* **1**, 175.

Norwegian Institute for Air Research: 1984, *Emission Sources in the Soviet Union,* 0–8147, Lillestrom, Norway, February, 29 pp.

Spangler, W. M. L. and Jenne, R. L.: 1985, *World Monthly Surface Station Climatology (and Associated Data Sets),* National Center for Atmospheric Research, Boulder, Colorado, May, 15 pp.

Szepesi, D. J. and Fekete, K. E.: 1987, *Atmos. Environ.* **21**, 1.

Summary Report from the Chemical Co-Ordinating Centre for the Second Phase of EMEP: 1983, *Cooperative Programme for Monitoring and Evaluation of the Long-Range Transmission of Air Pollutants in Europe,* EMEP/CCC-Report 4/83, November.

THE ASSESSMENT OF IMPACTS OF POSSIBLE CLIMATE CHANGES ON THE RESULTS OF THE IIASA RAINS SULFUR DEPOSITION MODEL IN EUROPE

S. E. PITOVRANOV

International Institute for Applied Systems Analysis (IIASA), A-2361 Laxenburg, Austria

(Received November 17, 1987; revised March 29, 1988)

Abstract. An analysis is made of the relationship between patterns in atmospheric circulation in Europe and the temperature regime of the Northern Hemisphere over the same period. The basis for classifying different types of atmospheric circulation or large-scale weather paterns [commonly known as Grosswettertypes (GWT-s) or Grosswetterlagen (GWL-n)] is the identification of the position of centers of cyclones, ridges and troughs. The linear regression between the frequency distribution of GWL-n and the deviation in the mean annual Northern Hemisphere extratropical temperatures from the 90-yr period (1891 to 1980) were tested. The results show that the null hypothesis, i.e. that there no linear relationship, is rejected at the 95% probability level (assuming a normal distribution) for several GWT-s and GWL-n. Changes in GWT-s and GWT-n frequency distribution associated with global warming could substantially change the long-range transport of pollutant over Europe. For example, the decrease in frequency of zonal circulation regimes and the more frequent meridional and blocked circulations (especially easterly flows) could result in a decrease of the existing net export of S pollutants from western to eastern Europe during the winter months.

1. Introduction

The International Institute for Applied Systems Analysis (IIASA) Regional Acidification Information and Simulation (RAINS) model attempts to provide scientific information for long-term strategies for controlling acidification impacts in Europe. The annual source-receptor matrix (SRM) is used in the model in forecasting S deposition in Europe as a function of different emission scenarios. The SRM has been constructed based on the results of the computations of 4 yr S transport and deposition in Europe for 1978 to 1982. These computations were made on long-range transport model of S pollution (Eliassen and Saltbones, 1983). The SRM was made available to IIASA by the European Monitoring and Evaluation Program (EMEP), Meteorologic Synthesizing Center-West in Oslo, Norway.

The time horizon of the RAINS model is 1960 to 2030. Many climatologists now believe that the global climate could noticeably change in the future 30 to 40 yr (see, for example, Bolin *et al.*, 1986). The main foundation of such forecasts is observed and projected increase in CO_2 and some other trace gases concentrations (CH_3, N_2O, freon, etc.) in the atmosphere as a result of anthropogenic activity. Associated with these changes a global warming due to the 'greenhouse effect' in the atmosphere could reflect on the conclusions of the RAINS model which is based on the standard SRM for 1978 to 1982.

The character of the climatic impact by an increase in the concentrations of CO_2 and other trace gases cannot be viewed in terms of a rise in global temperatures only. In fact, changes will take place in atmospheric pressure patterns, both geographical and seasonal, will in turn affect rainfall, temperatures, winds and all other meteorological variables that contribute to the overall climate at a given place (Lough *et al.*, 1983). This means also that by changing the frequency of the atmospheric circulation patterns, the dispersion and deposition patterns of emitted SO_2 and other compounds will also change. The goal of this study is to examine various methods for determining features of climatic change relevant to the long range atmospheric transport of S, and to make a preliminary effort in applying these changes to the RAINS model.

2. Scenarios of Climatic Change

2.1. APPROACHES TO CONSTRUCT CLIMATIC CHANGES SCENARIOS

An evaluation of the large number of results from climate model leads to the conclusion that the global equilibrium temperature change expected from increases of CO_2 and other greenhouse gases from AD 1980 to 2030 is likely to be in the range 1.0 to 2.5 °C (Bolin *et al.*, 1986). Three major techniques are available for the development of climatic scenarios.

– Numerical models form the basis of the first technique employed in the development of climatic change scenarios. Only three-dimensional General Circulation Models (GCM) contain the physical detail and geographical resolution necessary for even regional impact assessment.

– In the second method, instrumental data (i.e., data that have been measured in the past for selected climate variables) are used to describe extreme warm or cold years or periods that have occurred during the time span for which instrumental records are available. Such extreme historical cases may be treated as scenarios for a future altered climate state.

– The third method for constructing climate scenarios involves the use of paleo-climatic data. This may be defined as indirect information (tree rings, ice cores, archeological data, etc.) in preinstrumental warm periods, searching for clues to the possible behavior of climate in a future warm state.

The specific advantages and disadvantages of these methods are very strongly linked to the nature of the questions posed in the initial stages of the investigations. For the purpose of this study, the two first methods for developing climatic scenarios were chosen.

2.2. THE GCM SCENARIOS

The output available from GCMs for the analysis of simulated climate includes the three-dimensional distribution of the same basic variables that are used as input initial conditions (Gates, 1985). These include the wind field and rates of precipitation, which also figured as input parameters for the EMEP models. The main deficiencies of GCMs as forecasting tools at present are unsatisfactory reliability in providing regional detail (especially in precipitation and wind fields), and their limitation to studying only the steady-state response for a given increase in CO_2. Nevertheless, the GCMs data have been used in various applied studies which used rather detailed climatic data [see, for example, U.S. EPA (1984), and Parry *et al.* (1988)].

As in the abovementioned studies, the model produced by Goddard Institute of Space Studies (GISS) has been chosen for this study. The GISS model's output was originally

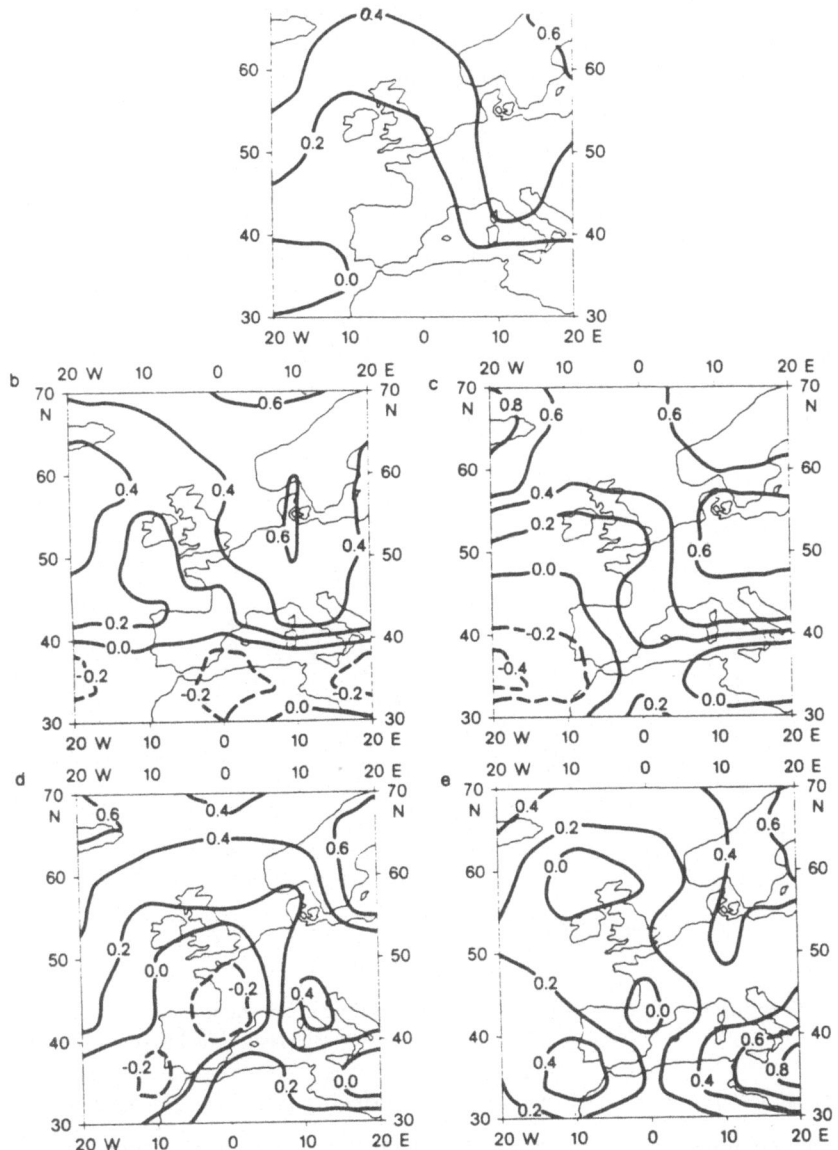

Fig. 1. Regional distribution of the precipitation of the rate change ($mm \cdot day^{-1}$) for this GISS $2 \times CO_2$ experiment. a: annual mean; b: winter; c: spring; d: summer; e: autumn (Bach, 1984).

provided for 8 × 10° grid squares. In order to provide additional spatial resolution, the
GISS output was modified by Bach (1984). A denser network of grid points was created,
thereby providing a resolution of 4° latitude × 5° longitude. Bach (1984) conducted
statistical analyses involving the testing of the statistical significance of the GCM results,
and the examination of model performance in simulating the features of present-day
climate on monthly basis. All comparisons, significance testing, and verification studies
were performed for the West European study area (20° W to 20° E and 30° to 70° N).
The study of Bach provided the same analysis and data for the whole EMEP grid area.
The maps of seasonal and annual precipitation fields for the GISS scenarios for a
doubling of CO_2 concentrations can be seen on Figure 1.

2.3. CLIMATE SCENARIOS BASED ON INSTRUMENTAL DATA

The range of variations in Northern Hemisphere temperatures during the last 100 years
is in the order of 0.5 °C (see Figure 2). A relatively dense network of meteorological
stations on land (especially in Europe) during this period provide the data for the
construction of various scenarios. For the purpose of this study two scenarios have been
chosen, one developed by Palutikof *et al.* (1984) and one by Vinnikov and Groisman,
(Vinnikov, 1986) on the basis of precipitation and surface pressure measurements in
Europe over the last 100 yr. In addition, scenarios have been developed based on the
frequency of occurrence of different types of atmospheric circulation in Europe.

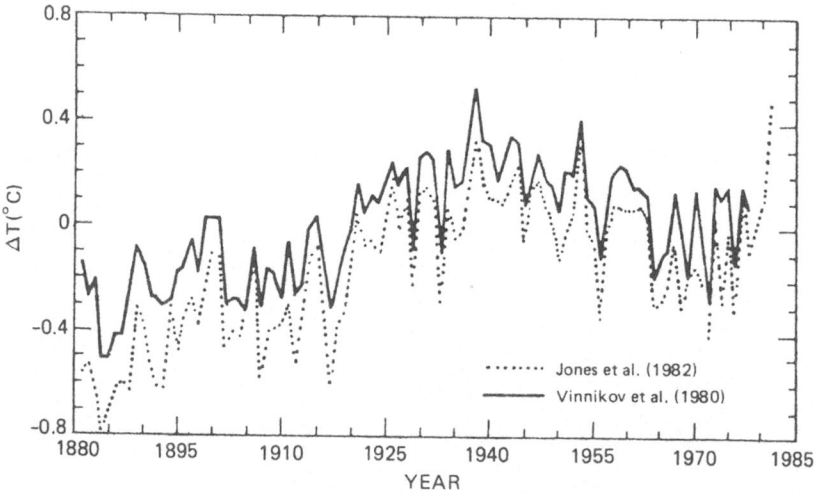

Fig. 2.   Comparison of the reconstructions of annual surface air temperature anomalies for the northern
hemisphere from Jones and Wigley (1982) and Vinnikov *et al.* (1980) (CDAC, 1983).

2.3.1. *A Scenario based on Comparison of Warmest and Coldest Decades*

Wigley and Jones (1981) looked at the signal-to-noise ratio of annual and seasonal
temperature series for the Northern Hemisphere temperature series in low-, mid-, and
high-latitude bands. They found that the annual Northern Hemisphere temperature

series had almost the highest signal-to-noise ratio. Lough *et al.* (1983) and Palutikof *et al.* (1984) choose the warmest and coldest sets of 20 consecutive yr from the average annual Northern Hemisphere temperatures for the period 1901 to 1980 inclusive. The warmest 20-yr period is 1934 to 1953, and the coldest 20-yr period is 1901 to 1920. The average annual hemispheric temperatures for these two periods are different by 0.4 °C. The maps of seasonal precipitation and sea level pressure differences in Europe between the warm period (1934 to 1953) and the cold period (1901 to 1920)) shown as Figures 3 and 4. Figures 5 and 6 show maps of changes in interannual temperature and precipitation variability for Europe (warm-minus-cold) from Palutikof *et al.* (1984).

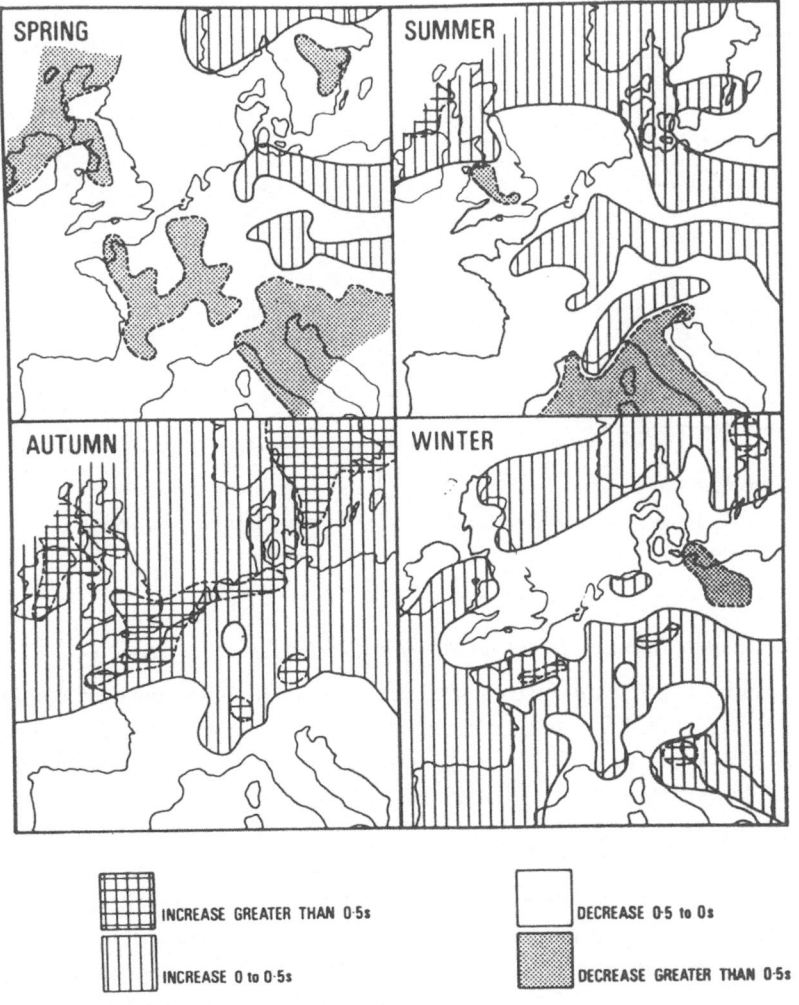

Fig. 3.   Precipitation change scenarios for Europe (Palutikof *et al.*, 1984).

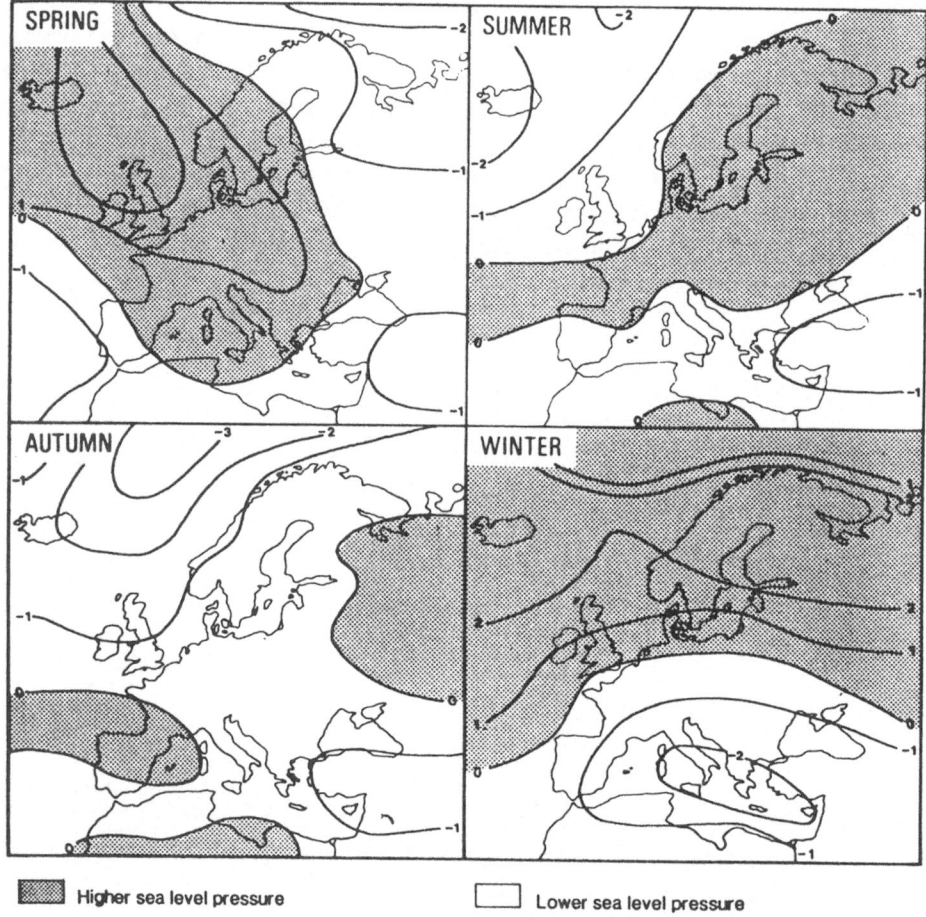

Fig. 4. Sea level pressure changes for Europe (Palutikof *et al.*, 1984).

2.3.2. *A Climate Scenario Based on the Vinnikov–Groisman Approach*

An alternative approach to instrumental scenario construction has been employed by Vinnikov *et al.* (1987). The most complete description of their approach and results is presented by Vinnikov (1986). He suggested a linear model for relating the values of regional climatic parameters to the hemispheric temperature

$$Y_i = \alpha_i \Delta T(t) + \beta_i + \varepsilon_i(t),\tag{1}$$

where t is the time, i is the designator for a local climatic characteristic, $\varepsilon_i(t)$ is a random model error, β_i is the bias parameter, α_i is the coefficient of proportionally between the local Y_i climatic variables and $\Delta T(t)$ (temperature for the extra-tropical part of the hemisphere in the zone 17.5 to 87.5° N). The parameters of such linear relationships have been assessed for the seasonal precipitation and surface atmospheric pressure in

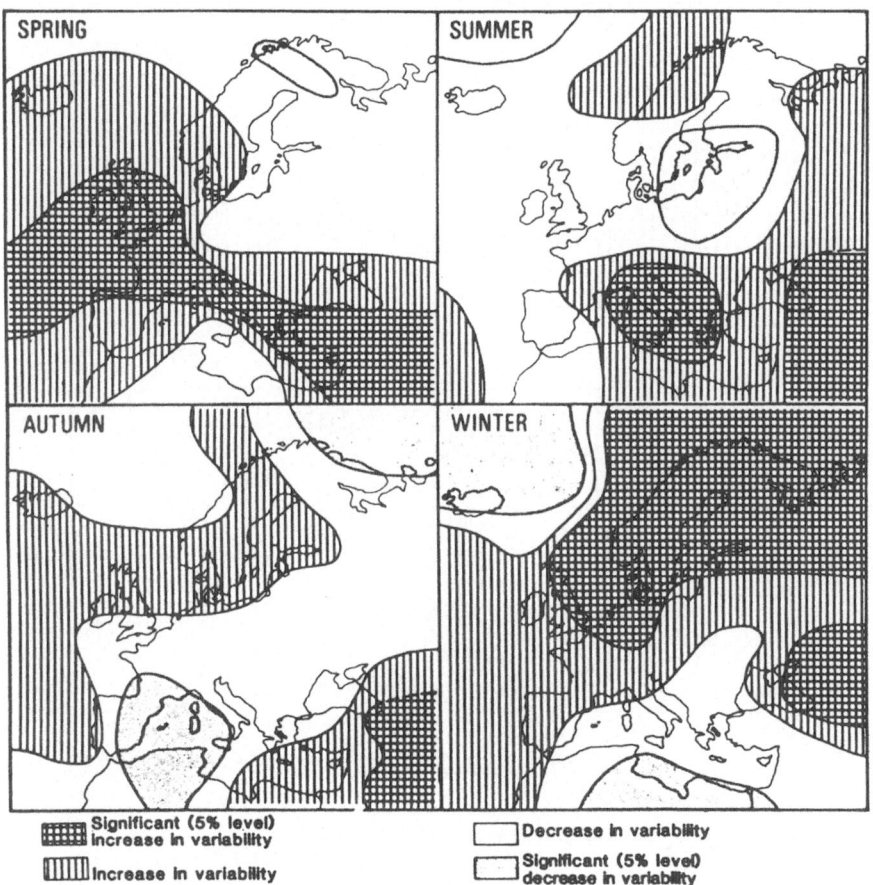

Fig. 5.   Changes in interannual temperature variability for Europe (warm-minus-cold) (Palutikof *et al.*, 1984).

the Northern Hemisphere (see Figure 7, after Kovyneva (1984) and Figure 8, from Vinnikov (1986)). Once α_i and ΔT have been determined, the change in precipitation and surface pressure for Europe can be estimated for any given temperature change; in this way regionally and seasonally specific scenarios can be developed. The Vinnikov–Groisman approach has some advantage in optimalizing all the information in a climatic time series, and does not rely simply on data covering some extreme years, which may be the result of random fluctuations in the climatic system. The statistical analysis employed permits the computation of important statistical characteristics such as confidence intervals of the parameters.

A comparison of the selected scenarios from Palutikof *et al.* (1984) and Vinnikov (1986) shows many similarities in suggested changes in climatic parameters, and close relationships between the spatial distributions of pressure, rainfall and temperatures in Europe. These findings are more evident when a warm climate is compared with a cold

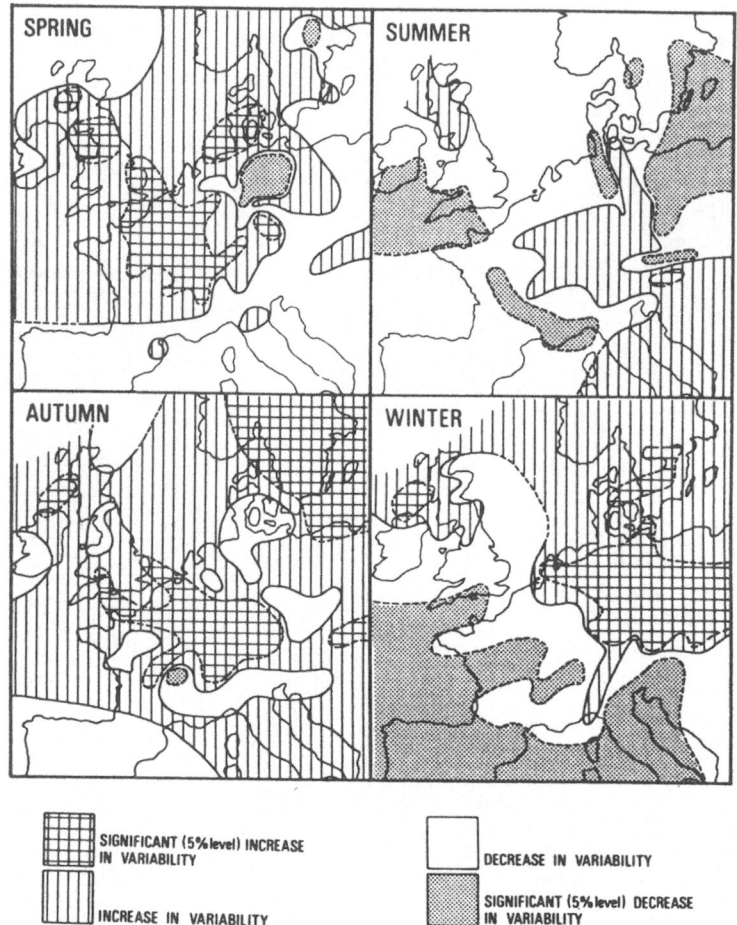

SIGNIFICANT (5% level) INCREASE IN VARIABILITY

INCREASE IN VARIABILITY

DECREASE IN VARIABILITY

SIGNIFICANT (5% level) DECREASE IN VARIABILITY

Fig. 6.   Changes in interannual temperature variability for Europe (warm-minus-cold) (Palutikof *et al.*, 1984).

one, but one should keep in mind that they were based on a change of the mean annual hemispheric temperature in the order of magnitude of $\pm 0.5\,°C$. For that reason, empirically-based scenarios 'can only be used in the early decade of the twenty-first century, after which, changes in boundary conditions and other no-analogue effects must reduce the relevance of past changes as analogue of the future' (Palutikof *et al.*, 1984).

3. Results and Discussion

3.1. THE STUDY OF DIRECT IMPACTS OF CHANGES IN PRECIPITATION

Some parameters of the scenarios of climatic change can be used as input parameters for the EMEP model to study the impacts of the expected changes in winds and

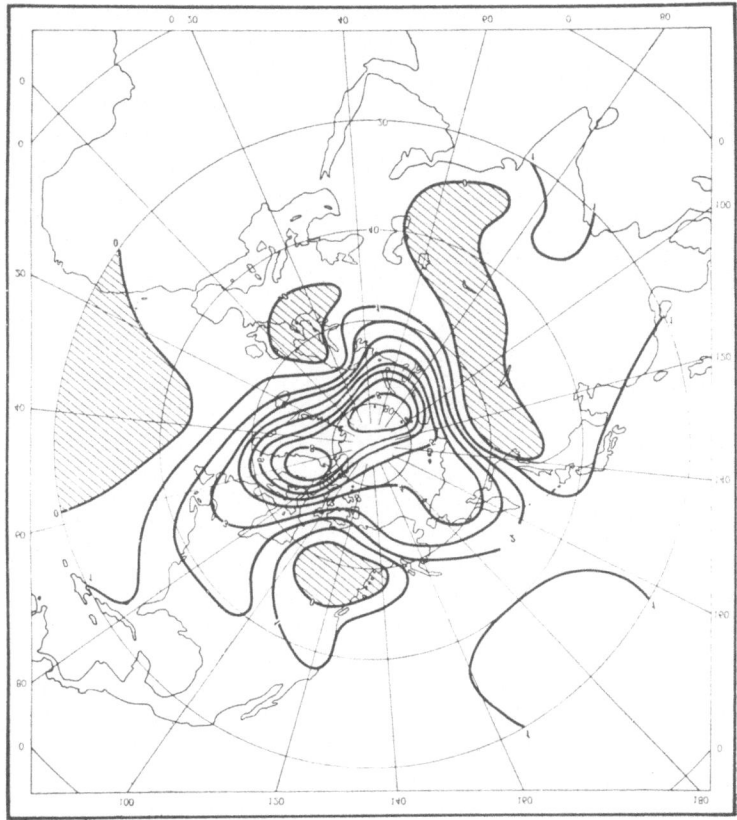

Fig. 7a.

Fig. 7.   Parameter estimates for the precipitation field, as percentage changes from the normal in seasonal totals per 0.1 °C increase in hemispheric temperature (winter – a, spring – b, summer – c, autumn – d) (Kovyneva, 1984).

precipitation patterns on the transport and deposition of SO_2 and SO_4 in Europe. However, the scenarios of climatic change have substantially coarser spatial and, in particular, temporal resolutions than the EMEP input data. Therefore, it is necessary to develop means of rendering scenarios of climatic change compatible with the EMEP input data.

The following two experiments with the EMEP model have been executed. Two source-receptor relations have been chosen for the analysis. These relations are: GDR-Illmitz (Austria) and UK-Rorvick (Sweden), (Alcamo and Bartnicki, 1985). The trajectories for these source-receptor relations have been calculated with the real wind data of 1980. Precipitation data for 1980 have been used for the calculation of the concentrations and depositions of SO_2 and SO_4 along these trajectories. This scenario is defined as the basic scenario.

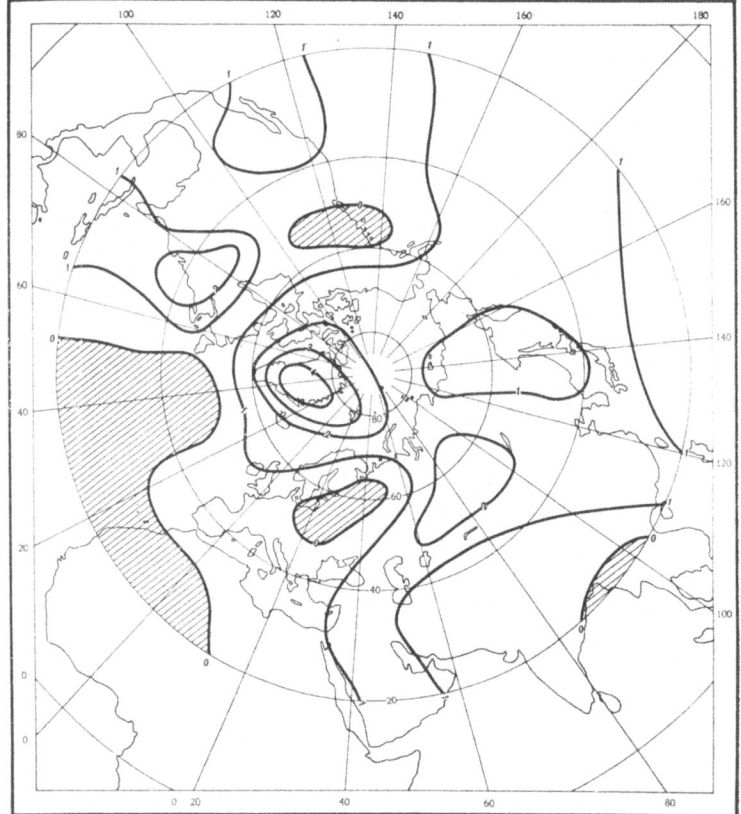

Fig. 7b.

An important assumption is that the trajectories for each of these two source-receptor relations will not change, but that the precipitation patterns will vary according to the scenarios of a climatic change. The EMEP model is insensitive to the precipitation in amount; its results are influenced only by the precipitation duration.

The simplest assumption of a uniform distribution of seasonal precipitation changes through each day of the season has been used. These values have been added to the observed daily precipitation data for 1980. These changing precipitation fields have been used as input parameters for the EMEP model.

Three scenarios of precipitation changes have been studied: GISS $2 \times CO_2$, Palutikof (Palutikof *et al.*, 1984), and Vinnikov–Groisman (Vinnikov, 1986). The results of the computations for the two source-receptor relations are shown in Tables I and II.

It can be seen from Table I, that the most sensitive state variables in the EMEP model to precipitation changes for the source-receptor combination GDR-Illmitz are dry and wet S depositions. The changes in dry and wet deposition have opposite signs, so the total depostion change in the receptor point is changing relatively small (10% from the basic scenario for GISS $2 \times CO_2$ scenario). The reason for the opposite changes in wet

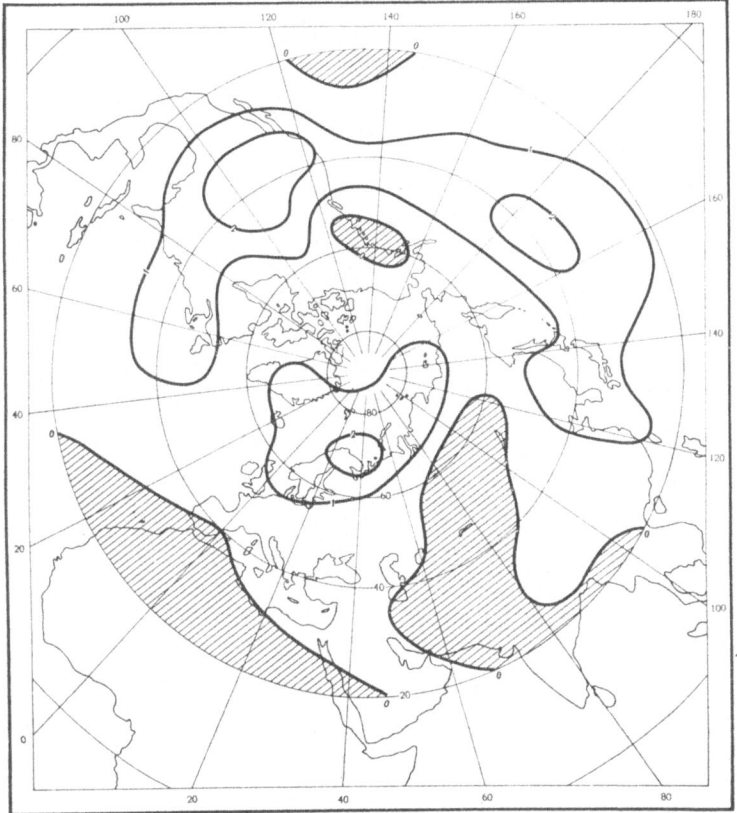

Fig. 7c.

and dry deposition is that according to all three scenarios, the mean annual of precipitation is increasing in the areas along most trajectories.

For the UK–Rorvick source-receptor combination the trajectories have passed over the sea. For these trajectories, the main mechanism of S removal from the atmosphere is wet deposition, which is close to its limit (saturation). Therefore, the increased precipitation amount in the area (also according to all scenarios) resulted only in changes of surface SO_2 and SO_4 concentrations. The dry, wet, and total deposition at the receptor point are practically unchanged.

Summarizing, it is possible to conclude that the results of the EMEP model are quite insensitive to changes in precipitation fields which could be associated with global warming. However, it is impossible to conclude that global warming could not influence the long-distance pollution deposition in Europe for two reasons. The first one is the possible deficiency of the EMEP model in describing the mechanism of wet S removal. The EMEP 2 version has a more sophisticated simulation of wet S removal from the atmosphere. The second reason may be the unjustified assumption that the pattern of trajectories is unaffected by climatic change.

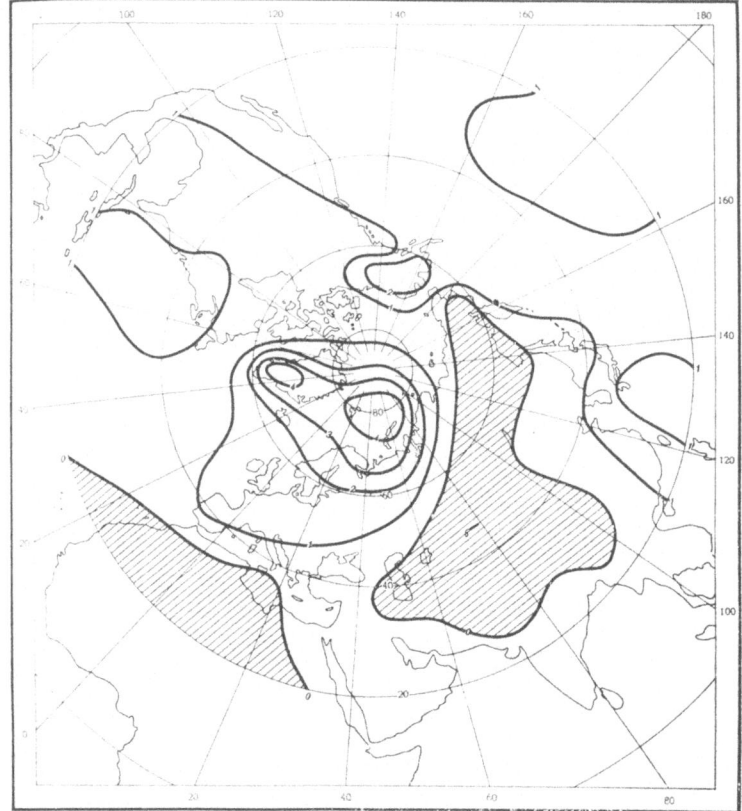

Fig. 7d.

The empirical scenarios and GCM models show that global warming result in changes in pressure patterns over Europe. These changes should result in changes in trajectories. It is possible in principle to assess the shift in geostrophical winds at the level 850 mb by the empirical maps of changes in surface seasonal pressure patterns in Europe (see Section 2.3). However, the uncertainty related to the transformation data and the EMEP input data could be very great. Therefore, in the following section, another approach which coincides better with the goals of this study, will be used.

3.2. THE RELATIONSHIP BETWEEN HEMISPHERIC TEMPERATURE AND WEATHER TYPES IN EUROPE

The basis for classifying different types of atmospheric circulation or large-scale weather patterns, commonly known as Grosswetterlagen (GWL-n), is the identification of the arrangements of the features of atmospheric pressure, such as anticyclones and cyclones, ridges, and troughs. On the basis of the work by Bauer (1947), Hess and Brezovsky (1952) distinguished 30 GWL-n with respect to Western Europe. Similar GWL-n were combined in general circulation types (GWTs) (Figure 9). For discussions

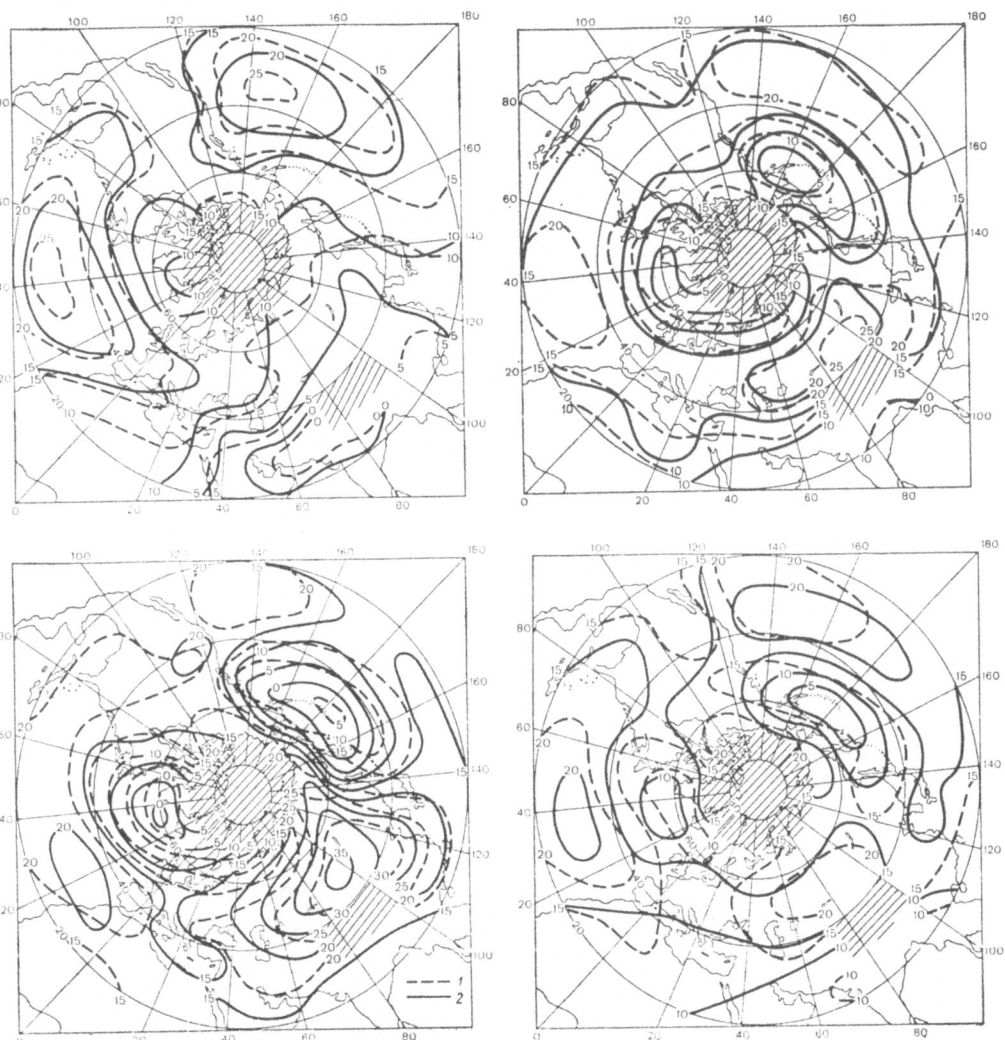

Fig. 8.   The sea level pressure (actual pressure minus 1000 KPA) (1) cooling − 0.5 °C, and (2) warming + 0.5 °C (winter – a, spring – b, summer – c, autumn – d) (Vinnikov, 1986).

about the behavior of the fundamental characteristics of atmospheric circulation, a final simplified classification results in only three general circulation regimes (GWRs). A detailed description and notation of GWL-n, GWT-s, and GWRs are given by den Tonkelaar (1988).

It is natural to expect that the observed changes in regional climatic parameters which accompany the global cooling and warming have to be reflected in frequencies of different types of European atmospheric circulation. The analysis of the existence of this relationship will be the goal of the following statistical study of frequency distribution of GWL-n, GWT-s, and GWRs over the last 90-yr period.

TABLE I

The impact of scenarios of precipitation changes on EMEP state variables [GDR-Illmitz (Austria)]

State variables	Basic scenario	Palutikof scenario	Vinnikov scenario	GISS $(2 \times CO_2)$ scenario	Maximum deviation from basic scenario (in %)
SO_2 (air) ($\mu g\,m^{-3}$)	1.02	0.93	0.91	0.87	16
SO_4 (air) ($\mu g\,m^{-3}$)	0.87	0.77	0.76	0.73	16
Dry deposition ($g\,m^{-2}\,yr^{-1}$)	0.33	0.28	0.26	0.20	44
Wet deposition ($g\,m^{-2}\,yr^{-1}$)	0.25	0.28	0.30	0.33	40
Total deposition ($g\,m^{-2}\,yr^{-1}$)	0.58	0.56	0.56	0.53	8

TABLE II

The impact of scenarios of precipitation changes on EMEP state variables [UK-Rorvick (Sweden)]

State variables	Basic scenario	Palutikof scenario	Vinnikov scenario	GISS $(2 \times CO_2)$ scenario	Maximum deviation from basic scenario (in %)
SO_2 (air) ($\mu g\,m^{-3}$)	0.23	0.22	0.17	0.14	30
SO_4 (air) ($\mu g\,m^{-3}$)	0.34	0.33	0.28	0.22	30
Dry deposition ($g\,m^{-2}\,yr^{-1}$)	0.065	0.065	0.062	0.062	0.5
Wet deposition ($g\,m^{-2}\,yr^{-1}$)	0.27	0.27	0.27	0.27	0
Total deposition ($g\,m^{-2}\,yr^{-1}$)	0.33	0.33	0.33	0.33	0

As shown in different studies, (see, for example, Palutikof *et al.* (1984) and Vinnikov (1986)), the characteristic changes of climatic parameters for warm and cold world depend upon the different seasons of the year. Therefore, the changes in the frequency of GWL, GWT, and GWR per season [winter (D, J, F), spring (M, A, M) summer (J, J, A) and autumn (S, O, N)] are analyzed in the present paper. Information on the daily distribution of the GWL for the region extending from 30 to 70° N and 40° W to 40° E, for the period 1891 to 1980 has been supplied in catalogues by the Koninklijk Nederlands Meteorologisch Instituut. Daily GWL data for the period 1891 to 1948 were determined on the basis of surface pressure data. In addition for the later period (1949 to 1980), 500 mb geopotential height data were used. Mean annual temperature data for the Northern Hemisphere over the period 1891 to 1976 was obtained from Gruza and Ran'kova (1979), and for 1977 to 1980 from Vinnikov *et al.* (1987).

3.2.1. *The Comparison of Changes of GWL and GWT During Warmest and Coldest Periods*

For mutual comparison of seasonal GWL, GWT, and GWR frequency distributions the period 1901 to 1920 was chosen as the coldest and 1934 to 1954 as the warmest 20-yr periods of the century, following the study of Palutikof *et al.* (1984). The statistical

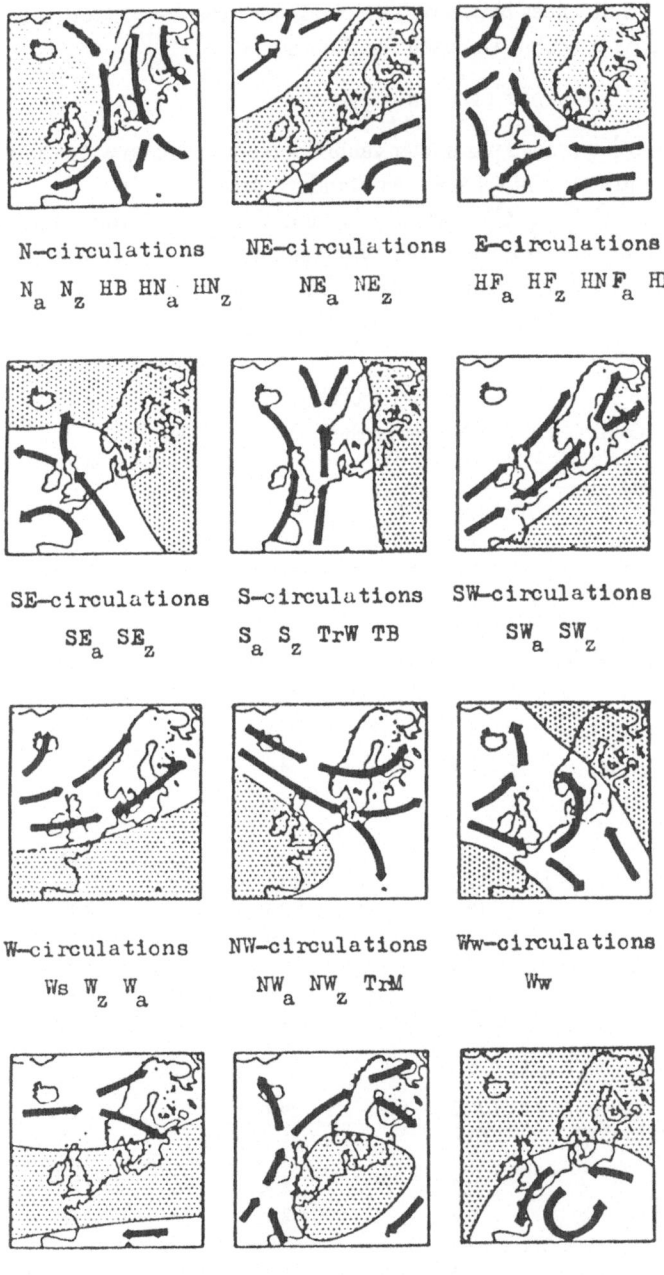

Fig. 9.   Schematic classification of the dominant airflows over Western Europe. In the dotted areas the pressure is high. The arrows are mainly related to the layers above 850 mbar (den Tonkelaar, 1988).

significance of differences between these two 20-yr samples was tested using the t-test. The following assessment was used for the t statistic:

$$t = (\bar{v}_1 - \bar{v}_2)/(\sigma^2(v_1) + \sigma^2(v_2)), \tag{2}$$

where $\bar{v}_1$, $\bar{v}_2$ are 20 yr mean parameter values, corresponding to warm and cold periods, respectively, and $\sigma(v_1)$, $\sigma(v_2)$ are their standard deviations.

Table III includes the GWLn, GWTs, and GWRs for which the hypothesis of absence of differences between the two samples is rejected with the probability exceeding 0.6 (under the assumption of normal distribution).

TABLE III

The frequency of GWL, GWT, and GWR for which the hypothesis of an absence of differences is rejected, with a probability exceeding 0.6 (under assumption of normal distribution)

Circulation	1901–1920 (T = 7.15 °C)	1934–1953 (T = 7.55 °C)	t-test t	Probability level
Winter				
Zonal GWR	26.6	21.7	− 0.4	> 0.6
Meridional/blocked GWR	33.4	41.7	0.45	> 0.6
High C.E. GWT	17.4	13.7	0.27	> 0.6
Easterly GWT	6.1	9.6	0.32	> 0.6
W_z (3) GWL	15.6	12.3	− 0.34	> 0.6
HM (5) GWL	13.6	8.6	− 0.39	> 0.6
Spring				
NW-ly GWT	8.0	5.5	− 0.65	> 0.6
Northerly GWT	12.8	19.1	1.8	> 0.95
Easterly GWT	10.3	8.5	− 0.36	> 0.6
HNa (10) GWL	2.9	5.4	0.57	> 0.6
HB (12) GWL	3.1	6.0	0.67	> 0.6
Ww (28) GWL	2.1	1.1	− 0.37	> 0.6
u (29) GWL	1.0	0.5	− 0.35	> 0.6
Summer				
Mixed GWR	26.2	29.4	0.26	> 0.6
Meridional/blocked GWR	41.2	36.3	− 0.38	> 0.6
High C.E. GWT	12.9	16.3	0.33	> 0.6
SW-ly GWT	0.2	2.3	0.6	> 0.6
Northerly GWT	16.9	8.6	− 0.95	> 0.75
NWa (8) GWL	9.6	5.6	− 0.44	> 0.6
N_z (14) GWL	5.4	0.4	− 0.85	> 0.75
Autumn				
High C.E. GWT	20.0	17.0	− 0.43	> 0.6
SW-ly GWT	2.8	5.0	0.73	> 0.75
Wz (3) GWL	10.7	14.5	− 0.46	> 0.6
Ww (28)	1.8	3.5	0.46	> 0.6

The results in Table III suggest the following tendencies in the circulation frequency of the two periods. In winter, hemispheric warming is associated with a decreased occurrence of zonal and mixed circulations in relation to the corresponding increase in frequency of meridional circulation types mainly associated with colder winter weather. In contrast, in summer, the warm 20-yr period exhibits more frequent mixed circulation regimes but less frequent meridional circulations than in the cool period. There is also a noticeable decrease in the frequency of northern and NW-en circulation types. This decrease is associated with higher summer temperatures.

For the transient seasons, the differences in circulation regimes is less marked, although there are noticeable changes in some GWTs. For example, the Northerly circulation type, which in Europe during the spring is associated with cold and wet weather conditions, is more frequent at the approximate 95% probability level in the warm world period than in the cool period. Southwesterly flows are somewhat more frequent in autumn in warm world, at the 75% probability level. However, the t-test for two 20-yr samples of GWL, GWT, and GWR shows that the level of confidence of rejection of 'null hypothesis' (which means no difference between two samples) is quite low. Except for Northerly GWT in spring, the confidence is less than 95%. Therefore, the observed differences can simply be the results of natural fluctuations of circulation frequency around long-term mean values. Therefore, an additional statistical analysis has been done based on the 90-yr time series 1891 to 1980 of GWL, GWT, GWR, and the Northern hemisphere temperatures.

In a manner similar to the Vinnikov and Groisman approach, a linear relationship between the extra-tropical mean annual Northern Hemisphere temperatures and frequency of GWLs, GWTs, and GWRs is assumed. A linear regression procedure has been followed

$$v_i = \alpha_i \Delta T(t) + \bar{v}_i + \varepsilon_i(t), \tag{3}$$

where $\Delta T(t)$ is the deviation of the 5-yr running average of the mean annual Northern Hemisphere temperatures in the zone 30 to 85 °N from the 90-yr mean value (Gruza and Ran'kova, 1979; Vinnikov et al., 1987), $\bar{v}_i$ is 90-yr mean circulation frequency, and α_i is the coefficient of proportionality.

The coefficient of proportionality was estimated by least squares, and the significance of the coefficient was tested using the Student's t-test (see, for example, Draper and Smith, 1966). For t-values exceeding $|2|$ the 'null hypothesis' (i.e., that there is no linear relationship) is rejected at the 95% probability level (under the hypothesis of a normal distribution). The results of the computations for GWTs and GWRs are given in Table IV.

In the first two columns of Table IV values of 90-yr (1891 to 1980) mean and standard deviations of GWTs and GWRs for the four seasons are presented. The calculated coefficients of proportionality between hemispheric temperatures and GWT and GWR frequency are presented in the third column. The sign and values of the parameters represent the direction and value in GWT and GWR changes when hemispheric temperatures increase by 1 °C.

TABLE IV

The assessment of parameters of linear relationship between frequency of GWTs and GWRs and mean annual air surface temperature of extra-tropical part of the Northern Hemisphere (30 to 85° N).

Circulation	1891–1980 mean	1891–1980 st. dev.	α_i in (days/1 °C)	t-test	Probability (2 tailed)
Winter					
Zonal GWRs	**22.6**	**10.7**	**− 10.5**	**− 2.04**	**0.04**
Mixed GWRs	28.0	10.8	− 1.9	− 0.36	0.72
High C.E. GWTs	16.0	8.4	− 2.1	− 0.51	0.61
SW-ly GWTs	5.2	5.1	0.05	0.02	0.98
NW-ly GWTs	6.8	6.0	0.15	0.04	0.95
Meridional/ blocked GWRs	**39.1**	**13.4**	**12.6**	**1.98**	**0.06**
Northerly GWTs	8.5	7.2	2.0	0.57	0.57
Low C.E. GWTs	5.5	5.3	1.0	0.38	0.7
Southerly GWTs	7.2	5.1	− 0.7	− 0.29	0.75
SE-ly GWTs	4.4	5.2	2.2	0.92	0.37
Easterly GWTs	**7.4**	**7.2**	**7.1**	**2.03**	**0.05**
NE-ly GWTs	2.9	3.4	0.8	0.48	0.65
Ww GWTs	3.2	3.4	0.2	0.1	0.9
Spring					
Zonal GWRs	16.9	7.8	1.9	0.5	0.62
Transient GWL	**1.0**	**1.4**	**− 1.0**	**− 1.96**	**0.06**
Mixed GWRs	21.3	10.0	1.8	0.36	0.72
High C.E. GWTs	11.6	6.9	2.6	0.78	0.44
Sw-ly GWTs	3.2	4.0	1.2	0.62	0.54
Nw-ly GWTs	6.5	5.9	− 2.0	− 0.72	0.47
Meridional/ blocked GWRs	52.0	12.1	− 2.7	− 0.45	0.65
Northerly GWTs	**14.8**	**8.4**	**13.8**	**3.55**	**0.00**
Low C.E. GWTs	7.4	5.7	− 3.6	− 1.3	0.2
Southerly GWTs	7.6	5.4	− 3.1	− 1.2	0.25
SE-ly GWTs	4.4	4.2	1.1	0.5	0.62
Easterly GWTs	9.6	7.5	− 5.1	− 1.4	0.15
NE-ly GWTs	6.2	4.9	− 4.0	− 1.6	0.11
Summer					
Zonal GWRs	25.5	10.2	− 3.5	− 0.6	0.57
Mixed GWRs	**27.6**	**10.3**	**14.0**	**2.9**	**0.01**
High C.E. GWTs	**14.3**	**7.5**	**8.4**	**2.35**	**0.03**
Sw-ly GWTs	2.1	3.7	1.1	0.6	0.57
NW-ly GWTs	11.2	7.5	4.5	1.2	0.25
Meridional/ blocked GWRs	38.1	11.1	− 10.1	− 1.6	0.11
Northerly GWTs	**12.9**	**7.2**	**− 12.9**	**− 3.9**	**0.001**
Low C.E. GWTs	5.1	4.4	1.1	0.67	0.5
Southerly GWTs	6.6	5.7	3.0	1.1	0.27
SE-ly GWTs	0.5	1.5	0.3	0.42	0.66
Easterly GWTs	**5.7**	**5.7**	**− 5.7**	**− 2.1**	**0.05**
NE-ly GWTs	6.0	6.1	3.0	1.1	0.27
Ww GWTs	2.0	3.2	1.1	1.1	0.27

Table IV (continued)

Circulation	1891–1980 mean	1891–1980 st. dev.	α_i in (days/1 °C)	t-test	Probability (2 tailed)
Autumn					
Zonal GWRs	21.8	10.6	0.6	0.1	0.9
Mixed GWRs	28.9	9.0	− 5.5	− 1.1	0.27
High C.E.GWTs	18.4	8.4	− 5.5	− 1.2	0.25
SW-ly GWTs	4.4	4.3	− 0.9	− 0.4	0.69
NW-ly GWTs	6.1	4.5	0.9	0.4	0.69
Meridional/ blocked GWRs	38.9	12.1	4.9	0.86	0.4
Northerly GWTs	9.1	7.1	1.4	0.41	0.68
Low C.E. GWTs	5.7	5.5	1.3	0.47	0.64
Southerly GWTs	9.6	6.9	− 0.8	− 0.2	0.85
SE-ly GWTs	3.6	4.3	3.3	1.6	0.11
Easterly GWTs	5.5	4.8	− 2.0	− 0.8	0.45
NE-ly GWTs	2.6	2.9	0.4	0.26	0.8
Ww GWTs	2.8	3.7	1.3	0.69	0.5

The t-test for the parameters of linear relationship and two-tailed probabilities of the existence of linear relationship being rejected are given in the last two columns.

The GWT and GWR which correlate with the hemispheric temperature change (with probability level exceeding 95%) are indicated. From Table IV it follows that, on the 95% probability level, some relationships between changes in hemispheric temperatures and frequency of soime GWTs and GWRs do exist, as follows:

(a) *Winter*

– A negative correlation of zonal circulation regime with the temperatures in winter seasons. This type of circulation represents the normal eastward progression of the Rossby waves in the upper westerlies at middle latitudes. A global warming would be associated with a decreasing frequency of this normal progression of Rossby waves.

– A positive correlation between temperature and the frequency of meridional circulations over Europe in winter seasons, and an increased frequency of blocked circulation with higher temperatures.

– A positive correlation of Easterly GWTs and hemispheric temperature in winter.

(b) *Summer*

– In summer, variations in hemispheric temperatures are positively correlated with the frequency of mixed circulations and are negatively correlated with those of meridional circulation. Therefore, an increase in hemispheric temperatures accompanies a decrease of blocking in Europe, as a whole during the summer months. It should be mentioned, however, that the blocking GWL (characterized by a zonal circulation type, a southerly flow in Eastern Europe and a blocking anticyclone in the European part of the U.S.S.R.)

has a tendency to increase in frequency during warm periods in the Northern Hemisphere, though the strength of this relationship is not sufficient to reject the null hypothesis at the 95% level.

– Hemispheric temperatures in summer are positively related to the frequency of GWT High C.E. (characterized by ridges of high pressure or characterized by anticyclonic conditions over Central Europe).

– With a global warming, the frequency of the Easterly airstream over Europe decreases in the summer months.

– There is a strong relationship between an increase in hemispheric temperature and a decrease of the frequency of the northerly airstream in the summer months in Europe, with higher frequencies being associated with lower temperatures.

(c) *Spring*

– In spring, in contrast to summer, the frequency of GWT (N) is positively correlated with hemispheric temperature. The high correlation with the temperature is also observed for the so-called transient circulation type in spring.

(d) *Autumn*

– There is no significant relationship between variations in hemispheric temperatures and frequencies of circulation-types in the autumn months.

The conclusions of this statistical study can be compared with some conclusions presented by the Palutikof *et al.* (1984) and the Vinnikov–Groisman (Vinnikov, 1986) studies. The most surprising result of the empirical study by Lough *et al.* (1983), Palutikof *et al.* (1984), and Vinnikov and Groisman (Vinnikov, 1986) is that winter temperatures over a substantial part of Europe are lower and show greater interannual variability during the warm period than during the cool period. This was explained as being probably the result of increased blocking (Wigley *et al.*, 1986). The results of the statistical study of GWL, GWT, and GWR over Europe in this paper support this explanation. The decrease in the frequency of zonal circulation frequencies and the increase of blocking events in winter during global warming is associated with increased frequency of polar outbreaks and easterly flows, bringing cold air masses into Western Europe.

This conclusion also coincides with the result of Bates and Meehl (1986). In this paper, the effect of CO_2 concentrations on the frequency of blocking events in NCAR GCM (coupled to a simple mixed layer ocean model) has been analyzed. The model results show that a doubling of CO_2 would result in an increase in the mean 500 mb height in winter over Northern Europe and the North Pacific compared to a control scenario. The 500 mb height would also increased in summer, but to a smaller degree.

In spring, cooler regions in Central Europe which coincide with the warm period of Palutikof *et al.* (1984) and Vinnikov and Groisman (Vinnikov, 1986) can also be associated with an increased frequency of the Northerly type blocking circulations. The summer decreased meridional and increased mixed circulation frequencies in summer, as well as the strong decreases of the Northern circulation type, also coincides qualitatively with the Palutikof *et al.* (1984) and Vinnikov and Groisman (Vinnikov, 1986) conclusion that summers are warmer and drier in Western Europe in the warm world.

It should be mentioned that, for the transient seasons (spring and autumn), the statistical analysis reveals much weaker correlations between hemispheric temperatures and circulation patterns. This result qualifies the general assertion of Palutikof *et al.* (1984) that there are considerable differences of temperature and precipitation fields in Europe between warm and cold periods, and especially so for autumn. The identification of statistically significant relationships between hemispheric temperatures and circulation patterns confirms the worth of approaches that use hemispheric temperatures as the predictor for the construction of scenarios of regional climatic change in Europe.

3.3. CLIMATIC SCENARIOS BASED ON 'GWL APPROACH'

The idea of a climatic scenario based on the GWL approach can be formulated as follows. The first step is the construction of a warm world scenario in terms of the frequency distributions of GWRs and GWTs for each season. The second step is a search in the observed past for a season with a similar frequency distribution as the warm world scenario.

Once the parameters of linear relationships have been determined (see Table IV), the change in the frequencies of GWTs and GWRs can be estimated from Equation (3) for any given hemispheric temperature change and therefore, seasonally specific scenarios can be developed. As mentioned in Section 2, the increasing equilibrium temperature expected up to 2030 is likely to be in the range 1.0 to 2.0 °C (Bolin *et al.*, 1986). The problem of the thermal inertia of climatic systems is very disputable. According to some assessments (see, for example, Schlesinger, 1985) the thermal inertia of the World Ocean resulted in a 50 to 60% reduction of transient temperatures compared to the equilibrium temperature. Therefore, a 1 °C increase of hemispheric temperature is assumed for the scenario of global warming up to 2030.

The parameters of linear relationship between hemispheric temperatures and GWTs and GWRs frequency presented in Table IV have the sign and value of the parameters representing the direction and value in GWTs and GWRs changes, when hemispheric temperature increase above the assumed 1 °C. Therefore, the scenarios of the frequency distribution of GWTs and GWRs corresponding to 1 °C warmer world represent the sum of the mean 1891 to 1980 frequencies of GWTs and GWRs (the first column in Table IV) and parameters of linear relationships (the third column in Table IV).

There is a low probability of finding seasons in the past with exactly the same frequencies of GWTs and GWRs, due to the relatively short time series of observed GWTs and GWRs in the past. It can be seen from Table IV, that only a few GWTs and GWRs are significantly correlated with the hemispheric temperatures. Therefore, the least square procedure with case weights specified (with weights proportional to t-assessment) have been used in the procedure of searching for season-analogues.

$$\{v_1^{an}, \ldots, v_{13}^{an}\} = \min_{1 \le j \le 90} \sum_{i=1}^{13} t_i (v_i^{sc} - v_i^j)^2 \qquad (4)$$

where v_i^j and v_i^{sc} are the frequencies of ith GWT and GWR for the jth year and scenarios frequency, respectively, t_i is the value of the t-test, and $v_1^{an} \ldots v_{13}^{an}$ is the frequency distribution of the selection of minimization problem.

The results of the computations showed, that the most similar frequency dustributions of observed GWTs and GWRs to a warm world scenario distributions were observed in the following seasons: For winter 1938, spring 1939, summer 1952, and autumn 1939.

Three scenarios of frequency distributions of GWTs and GWRs are presented now: (1) for a warm world; (2) during the assessed season-analogues in the past, and (3) during the 4-yr October 1978–September 1982 of the standard source-receptor matrix period (Table V) for the EMEP long range atmospheric transport model.

TABLE V

The scenario of frequency distribution of GWT and GWR for the warm world, during the assessed season-analogues in the past, and during 4 yr (October 1978–September 1982) of the standard source-receptor matrix period

Circulation	Winter (analogue is 1938)			Spring (analogue is 1939)			Summer (analogue is 1952)			Autumn (analogue is 1939)		
	St.	Sc.	An.	St.	Sc.	An.	St.	Sc.	An.	St.	Sc.	An.
Zonal	**29.8**	**12.1**	**12**	12.2	18.8	19	27.2	22.0	22	26.8	22.4	14
Mixed	27.6	26.1	29	26.8	22.9	33	**18.3**	**41.6**	**45**	37.1	23.4	23
High C.E.	15.3	13.9	16	12.8	14.2	16	**12.8**	**22.7**	**19**	24.9	12.9	12
SW-ly	5.0	5.2	0	7.5	4.4	1	2.0	3.2	7	7.7	3.5	10
NW-ly	7.3	7.0	13	6.5	4.5	21	3.5	15.7	19	4.5	7.0	1
Merid./block	32.1	51.7	48	51.2	49.4	36	45.0	28.1	25	26.1	43.8	54
Northerly	6.0	10.5	14	**11.5**	**28.6**	**30**	**16.5**	**0.0**	**0**	3.0	10.5	19
Low C.E.	1.0	6.5	0	6.8	3.8	0	2.5	6.2	6	4.2	7.0	9
Southerly	7.8	6.5	6	15.7	4.5	0	8.5	9.6	11	8.0	8.8	1
SE-ly	7.5	6.6	3	3.7	5.5	0	2.3	0.8	0	1.5	6.9	12
Easterly	**7.8**	**14.5**	**13**	9.3	4.6	0	**11.2**	**0.0**	**4**	5.8	3.5	6
NE-ly	0.5	3.7	11	4.2	2.2	6	3.0	9.0	4	2.8	3.0	6
Ww	1.5	3.4	0	0.0	0.2	0	1.0	3.1	0	0.8	4.1	1

It can be seen from Table V, that the frequencies of all GWTs and GWRs which are significantly correlated with the hemispheric temperatures are in good agreement with the frequencies of season-analogues. For the winter seasons the maximal deviation of GWT and GWR from the warm world scenario frequencies occurs for the meridional circulation and is equal to 3.7 days. The frequency deviation of Northerly GWT in spring is 1.4 days. The maximal deviation in summer equal to 4 days occurs for Easterly GWT.

Also, it can also be seen from Table V, that the GWT and GWR frequency distributions for the source receptor matrix (SRM) period differs considerably from the supposed scenario. For example, the difference in the frequency of zonal circulations frequency in winter is 18 days, or 60% of the deviation. The difference in the frequency of Northerly GWT's in spring is 18 days, or 160% of the deviation. The difference in the frequency of mixed GWR's in summer is 27 days, or 150% of the deviation.

3.4. THE IMPLICATION OF 'GWL SCENARIO' FOR THE ASSESSMENT OF THE CLIMATIC IMPACT ON ACIDIFICATION PROCESSES IN EUROPE AND ACID RAIN CONTROL STRATEGIES

The results of the above statistical analysis show that the 'artificial year' which consists of winter 1938, summer 1952, spring and Autumn 1939 has the most similarity in terms of the circulation frequency distribution to the warm world scenario projected on the basis of a regressional analysis of the relationships between the frequency distributions of GWT's and GWR's, and the hemispheric temperatures over 90-yr period from 1891 to 1980. All seasons of the 'artificial year' belong to the warmest 20-yr period of the century. The mean annual air surface temperatures of he Northern Hemisphere in the latitude zone 30 to 87.5° N for 1938, 1952 and 1939 yr are equal to 8.0, 7.5, and 7.8 °C, respectively. The choice of the warmest individual years as an analogue of future warmer climate is a well-known approach in the construction of a climatic scenario (see, for example, Palutikof *et al.*, 1984). The non-trivial conclusion of the statistical analysis is that the most likely year (from the circulation point of view) is composed from several individual years. It should be mentioned, however, that 1952 does not belong to the warmest individual year of the century of the Northern Hemisphere.

It will be supposed hereafter, that the 'artificial year' could be applied to the analysis of climatic conditions in Europe at the early decade of the twenty-first century. The comparison of the GWT and GWR frequency distributions of the 'artificial year' and of the SRM years shows that significant differences in GWT and GWR frequencies could result in significant changes in predicted pollutant transport and deposition over Europe, which are for the time being based on the standard 4-yr period. On the qualitative level the main differences in transport and deposition will be as follows:

– The less frequent zonal circulation regime and more frequent meridional and blocked circulation (especially easterly flows) could result in a decrease of the existing and projected net export S pollutants from western to eastern Europe in the winter months.

– In spring, the pollutant transport from north to south in Europe could increase considerably, associated with more frequent northerly flows.

– In the summer, in contrast, the more frequent mixed circulations and less frequent meridional circulations may result in a decreasing pollutant transport from north to south in Europe. The lower frequency of easterly flows should decrease the pollutant transport in the east-west direction in summer. The high frequency of High C.E. GWT is associated with the increasing anticyclonic conditions over Central Europe which would result in a decreasing intensity of pollutant transport over Central Europe and a decreasing removal of contaminated pollutants from the atmosphere by precipitation.

Unfortunately, it is impossible to compare the observations of SO_2 and SO_4 concentrations and depositions for the seasons of the 'artificial year' with those during the SRM 4-yr period, because the regular measurements were started on this basis only during the

1970's as part of the EMEP Programme. However, it is possible to assess the significance of differences in circulation patterns between the 'artificial year' and the SRM standard period using a long-range transport model. It seems that several meteorological centers over the world have the archive weather infromation for the years 1938, 1939 and 1952 which could provide the input data for the models. Only a direct numerical experiment involving a very time-consuming computer run needed for obtaining the SRM could provide the background for the implication of the proposed climatic scenario for the acid rain control strategy in Europe.

4. Conclusions

A GWT and GWR (large-scale weather patterns) frequencies distribution scenario in 'a warm world', developed on the basis of a statistical analysis of relationships between the hemispheric temperatures and GWT and GWR frequency distributions, shows considerable differences with the GWT and GWR frequency distributions for the standard EMEP source-receptor matrix period (1978 to 1982). In particular, the decrease of zonal circulation in winter season could result in decreasing transport of pollutant from the western to eastern part of Europe in the event of a climatic warming.

A search of past climatic data for an annual analogue of GWT- and GWR-frequencies distributions similar to those for the warm world scenario resulted in an 'artificial year' composed of winter 1938, summer 1952, and spring and autumn 1939. The meteorological data available for these seasons could be used as input data for the EMEP model. The results of such an experiment could provide background information for the acid rain control strategy in Europe and incorporated possible climatic changes in the next few decades.

Acknowledgment

To J. den Tonkelaar for his idea to use GWL data for the construction of climate scenario for Europe, valuable discussions and providing the basic GWL data; to V. Fedorov for discussion about the statistical aspects of the study; to J. Alcamo, J. Jäger, and R. Munn for their valuable comments.

References

Alcamo, J. and Bartnicki, J.: 1985, 'An Approach to Uncertainty of a Long Range Air Pollutant Transport Model', *IIASA Working Paper WP-85–88*, International Institute for Applied Systems Analysis, Laxenburg, Austria.
Bach, W.: 1984, *Progress in Physical Geography* **8**, 583.
Bates, G. T. and Meehl, G.: 1986, *Mon. Weather Rev.* **114**, 687.
Bauer, F.: 1947, *Einfuhrung in die Grosswetterkunde*, Dietrich, Wiesbaden.
Bolin, B., Jäger, J., and Döös, B. R.: 1986, Ch. 1, 'The Greenhouse Efect, Climatic Change, and Ecosystems', in B. Bolin, B. R. Döös, and J. Jäger (eds.), *The Greenhouse Effects, Climatic Change, and Ecosystems*, SCOPE 29, John Wiley and Sons, pp. 1–32.

CDAC: 1983, *Changing Climate*, Report of the Carbon Dioxide Assessment Committee, National Academy Press, Washington, D.C.

Draper, N. R. and Smith, H.: 1966, *Applied Regressional Analysis*, John Wiley and Sons, Inc., New York.

Eliassen, A. and Saltbones, J.: 1983, *Atmos. Environ.* **17**, 1457.

Gates, W. L.: 1985, *Climatic Change* **7**, 267.

Gruza, G. V. and Ran'kova, E. Ya.: 1979, *The Data of the Structure and Variability of Climate. The Sea-level Temperatures. The Northern Hemisphere*, Obninsk, 203 pp.

Hess, P. and Brezowsky, H.: 1952, *Katalog der Grosswetterlagen Europas*, Ber. Dtsch. Wetterdienstes, US-Zone, No. 33, Bad Kissingen.

Jones, P. D. and Wigley, T. M. L.: 1982, *Climate Monitor* **9**, 43.

Kovyneva, N. P.: 1984, *Geographicheskaya* **6**, 29 (in Russian).

Lough, J. M., Wigley, T. M. L., and Palutikof, J. P.: 1983, *J. Clim. Appl. Meteorol.* **22**, 1673.

Palutikof, J. M., Wigley, T. M. L., and Lough, J. M.: 1984, *Seasonal Climate Scenarios for Europe and North America in High-CO$_2$, Warmer World*, U.S. Dept. of Energy, Carbon Dioxide Research Division, Technical Report TR012, 70 pp.

Parry, M. L., Carter, T. R., and Konijn, N. T. (eds.): 1988, *The Impact of Climatic Variations on Agriculture. Volume 1. Assessment In Cool Temperature and Cold Regions*, D. Reidel Publ. Co., Dordrecht, Holland.

Schlesinger, M. E.: 1985, *Equilibrium and Transient Effects of Increased Atmospheric CO$_2$*, Report No. 67, Oregon State University, Corvallis, Oregon 97331.

U.S. EPA: 1984, *Potential Climatic Impacts of Increasing Atmospheric CO$_2$ with Emphasis on Water Availability and Hydrology in the United States*, Strategic Studies Staff, Office of Policy Analysis, April.

Vinnikov, K. Ya., Gruza, G. V., Zakhazov, V. F., Kirillov, A. A., Kovyneva, N. P., and Ran'kova, E. Ya.: 1980, *Soviet Meteorol. Hydrol.* **6**, 1.

Vinnikov, K. Ya.: 1986, *Climatic Sensitivity*, Leningrad, Gidrometeoizdat, 223 pp. (in Russian).

Vinnikov, K. Ya., Groisman, P. Ya., Lugina, K. M., and Golubev, A. A.: 1987, *Meteorologia i Gidrologia* **1**, 45 (in Russian).

Wigley, T. M. L. and Jones, P. D.: 1981, *Nature* **292**, 205.

Wigley, T. M. L., Jones, P. D., and Kelly, P. M.: 1986, 'Empirical Climate Studies. Warm World Scenarios and the Detection of Climate Change Induced by Radiationly Active Gases', Ch. 6; in B. Bolin, B. R. Döös, J. Jäger, and R. A. Warrick (eds.), *The Greenhouse Effect, Climatic Change, and Ecosystems*, John Wiley and Sons, Chichester–NY–Brisbane–Toronto–Singapore, pp. 271–322.

FOREST CANOPY TRANSFORMATION OF ATMOSPHERIC DEPOSITION

MICHAEL BREDEMEIER

Research Centre Forest Ecosystems and Forest Decline, University of Göttingen, D-3400 Göttingen, West Germany

(Received November 17, 1987; revised April 27, 1988)

Abstract. Solute fluxes to the ground in open plots and under the forest canopy of different species were investigated in a number of long-term ecosystem studies in West Germany. From the canopy flux balance, rates of interception deposition and canopy/deposition interactions were assessed. Chemically, both open precipitation and throughfall are dilute solutions of H_2SO_4 and HNO_3 and their salts. For the sites investigated, mean pH in bulk precipitation ranged from 4.1 to 4.6, and in throughfall from 3.4 to 4.7. The increase in acidity after canopy passage at most sites indicates considerable interception deposition of strong acids to the forest stands, exceeding the rate of H^+ buffering in the canopy.

Evidence for buffering processes can be directly deduced from the fact that on sites with high soil alkalinity and high foliage base status, throughfall pH is usually higher than precipitation pH. Furthermore, the same idea can be concluded from changes in solution composition after canopy passage: the H^+/SO_4^{2-} ratio is decreasing at most sites, while alkali earth cations from exchange processes occur in throughfall (Ca^{2+}/SO_4^{2-} ratio increases). Solution composition and element flux data are presented for each of the sites, and the regional, orographical and site specific (species composition, ecosystem state) differentiations are discussed.

A method for the assessment of total deposition and of canopy interactions such as H^+-buffering and cation leaching is described, and results of calculations are shown. From these calculations it is concluded that forest ecosystems in Germany receive mean H^+ loads of ca. 1 to 4 keq $H^+ \cdot ha^{-1} \cdot a^{-1}$ from atmospheric deposition. Acidity deposition rates seem to be related to a few key factors such as regional characteristics and ecosystem characteristics.

1. Introduction

During the past decade, there has been much research on atmospheric depostion. Detailed knowledge has been achieved about wet and dry deposition processes under different micrometeorological conditions. Models of pollutant distribution, deposition and environmental impact have been formulated. But still, dry and total deposition rates to the terrestrial environment are very difficult to determine experimentally.

The reason is that 'physical' methods of dry deposition measurement usually require much experimental effort (air concentration gradients, eddy correlation) and have therefore only been pursued over relatively short periods of time (hours to days). In addition, the artificial sampling devices used in those methods might not represent the actual air pollutant capture characteristics of a real canopy. *Long-term* (e.g., annual) total wet plus dry deposition rates, which represent the input side in forest ecosystem element budgets, have been quantified by these techniques at only a very limited number of research sites (e.g., Walker Branch watershed, Lindberg *et al.*, 1986).

Solute fluxes in open land precipitation and forest throughfall, on the other hand, are easy to sample continuously over long periods of time, and this is done in several monitoring studies and wet deposition networks throughout the world. Fluxes in incident precipitation, however, are not equal to total deposition in most cases, because interception deposition of both gases and particles in dry form or of ions in fog and cloud water is not taken into account. Element transport in forest throughfall does not necessarily correspond to total deposition onto the canopy either, because canopy interactions with deposited constituents may occur.

Generally, these interactions appear as sink or source functions of the canopy for special chemical constituents. Examples for sink functions are the buffering of H^+ by cation exchange from inner leaf surfaces (Ulrich, 1983b; Meiwes and König, 1986; Roelofs et al., 1985), foliar nutrient uptake (Boynton, 1954; Matzner, 1986) and the retention (adsorption) of certain heavy metals on surfaces (Godt, 1986). Source functions of the canopy can occur as leaching of cations from exchange reactions (Ulrich 1983a, b) or from excretion, mostly in conjunction with organic anions (Hofman et al., 1980; Lutz, 1987).

For some consitutents, however, the canopy may act like an 'inert sampler', i.e., all dry deposited material is washed off the surfaces during rain events and appears in throughfall, while no significant foliar leaching/uptake would occur. For elements, that are deposited at very high rates compared to their possible internal cycling (e.g., S in Central Europe, see discussion below), it may be assumed that the canopy acts more or less in this way.

Thus, long-term averaged solute fluxes to the ground in forest stands may give an idea of the total deposition for some constituents. However, to assess deposition rates of elements which are involved in canopy interactions to a great extent, a method to *partition net throughfall* (i.e., flux in TF minus flux in open precipitation) into interception deposition and leaching/uptake is needed. Such a method, that rests on considerations of sources, binding forms and deposition mechanisms of major elements in throughfall, is proposed here. The calculation mode and its underlying assumptions will be explained, and results will be shown and discussed.

In addition, observed solute fluxes from a number of monitoring sites in Western Germany are reported, covering a wide range of geographical, orographical and ecological situations. These data were originally obtained and used for forest ecosystem budget studies (e.g., Ulrich, 1986; Matzner et al., 1984; Bredemeier, 1987; Büttner et al., 1986; Hauhs, 1985). They are also of value for comparison with results from regional deposition models.

It must be stated clearly, however, that the method for assessment of total deposition and canopy exchange described below is not a deterministic model, since it rests on assumptions that have no rigorous experimental proof for all the different meteorological and ecological situations occurring over long-term monitoring periods. The restrictions will be discussed with the explanation of the approach and in the general discussion.

2. Methods and Monitoring Sites

In ten forested ecosystems of northwestern Germany (Figure 1, Table I), samples of incident precipitation (above canopy or in open plots) and throughfall were collected weekly. These were all *bulk* samples from continuously open collectors. Thus, incident precipitation includes some portion of dry deposited coarse particles while throughfall includes cloud water drip from canopies at high elevation sites.

Weekly samples were stored at 4 °C and combined proportional to water amount for monthly samples, which were analyzed for the major ions: H^+ (pH), Na^+, K^+, Ca^{2+},

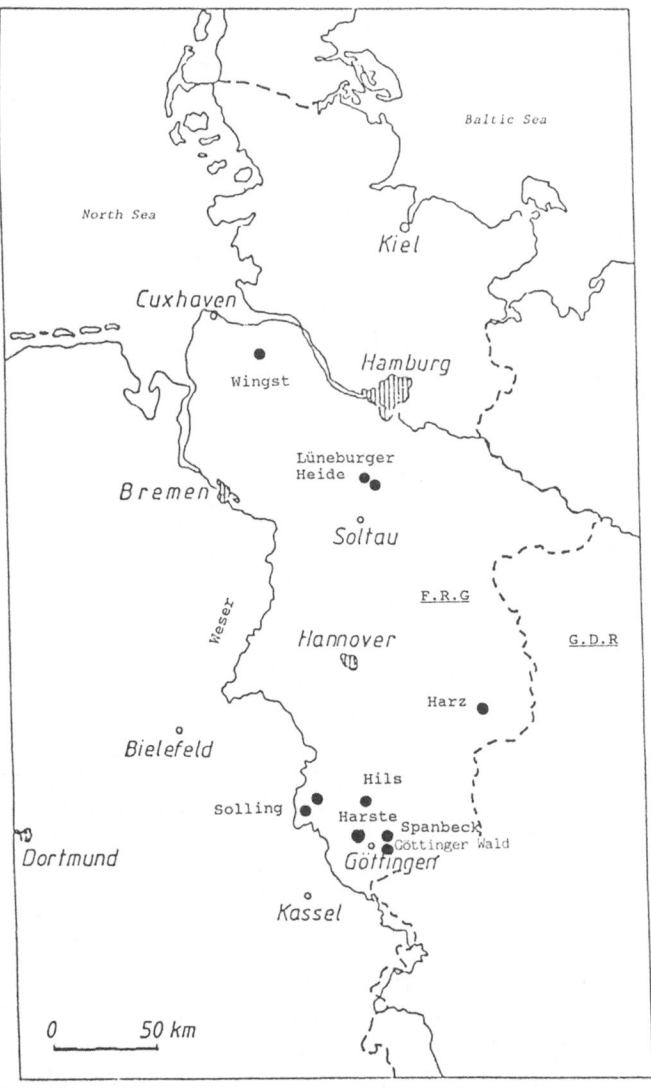

Fig. 1.   Locations of experimental forest sites in northwestern Germany.

TABLE I

Forest sites investigated in northwestern Germany

Site	Species	Age (yr)	Elev. (m)	Mean annual rain (mm)	Specific features
Wingst	spruce	95	40	870	Glacial sand, spodo-dystric cambisol; severely damaged, close to North Sea coast on exposed, elevated terrain
Lüneburger Heide	oak pine	105 100	90 80	800 800	Glacial loamy sand, spodo-dystric cambisol; no damage, rel. close to coast, sheltered in widespread forest area on flat plains
Harz	spruce	40	700	1270	Devonian sandstone, spodo-dystric cambisol; high altitude catchment, young and dense spruce stands, damaged (needle loss & yellowing)
Hils	spruce	100	460	1080	Hils-sandstone, podzol; exposed, damaged opening stand
Solling	beech spruce	138 103	500 500	1030 1030	Loess over sandstone (sm II), spodo-dystric cambisol; severely acidified soils, very exposed situation, spruce damaged
Harste	beech	96	200	610	Loess over limestone (mu), orthic luvisol; low degree of soil acidification, no damage
Spanbeck	spruce	86	250	600	Loess over sandstone (sm II), (spodo-)dystric cambisol; ongoing soil acidification, high growth rate, no visible damage
Göttinger Wald	beech	110	340	700	Limestone (mu), chromic cambisol; calcareous soil on limestone, throughfall alkalinity

Mg^{2+}, Fe^{3+}, Mn^{2+}, Al^{3+} (Atomic Absorption), NH_4^+, SO_4^{2-}, Cl^-, NO_3^-, PO_4-P (photometric auto analyzer). Phosphate analysis was mainly done for quality control, i.e., to detect contamination by birds, insects, etc. At each site, ten collectors were installed in the open, whereas 15 to 20 were located in the forest stands. Water from five samplers was combined into one replicate. Table I gives a short characterization of the investigated sites. They cover a range of different geographical/climatological, orographical and ecological situations in northwestern Germany. Elevations range from 40 m above sea level (Wingst) to 700 m in the upper Harz mountains. Mean annual rainfall varies from 650 to 1200 mm y^{-1}. Most of the forest sites consist of older mature stands. The species studied are the most important ones in West German forestry: spruce (*Picea abies L.*), beech (*Fagus silvatica L.*), oak (*Quercus robur L.*) and pine (*Pinus silvestris* L.). Some case studies (Lüneburger Heide, Solling, Harste/Spanbeck) are designed to compare element fluxes in deciduous and coniferous forest ecosystems. This is to investigate how species composition and ecosystem status affect deposition rates.

Some specific features of the sites, regarding their exposure, soil chemical conditions and visible damage symptoms, are listed in Table I. All spruce stands investigated show a high degree of soil acidification, many of them are more or less severely damaged (needle loss, yellowing, root dieback) (Streletzki, 1986). The deciduous sites seem to be generally less affected and usually show less soil acidification than the coniferous sites.

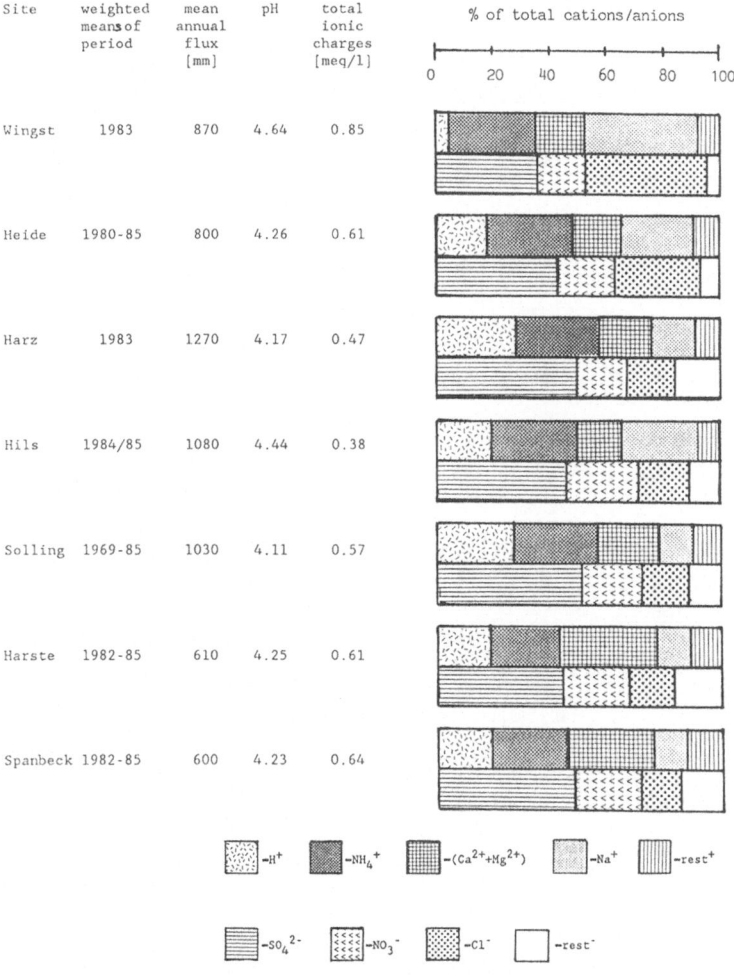

Fig. 2. Solution chemistry of open (bulk) precipitation.

3. Solution Composition of Bulk Precipitation and Throughfall

Some parameters to characterize solution chemistry in bulk rain and throughfall at a
selection of sites covering a transect from the North Sea shore to inner parts of the
country are given in Figures 2 and 3. All numbers are volume-weighted means of the
respective monitoring periods. The reader should keep in mind that the data being
compared here represent partly different time periods of different hydrologic characteristics. However, general trends in the data should still be apparent.

The range of pH in bulk precipitation is fairly narrow. For six of the seven open sites
in Figure 2, the weighted mean pH lies between 4.1 and 4.4. Only Wingst shows a

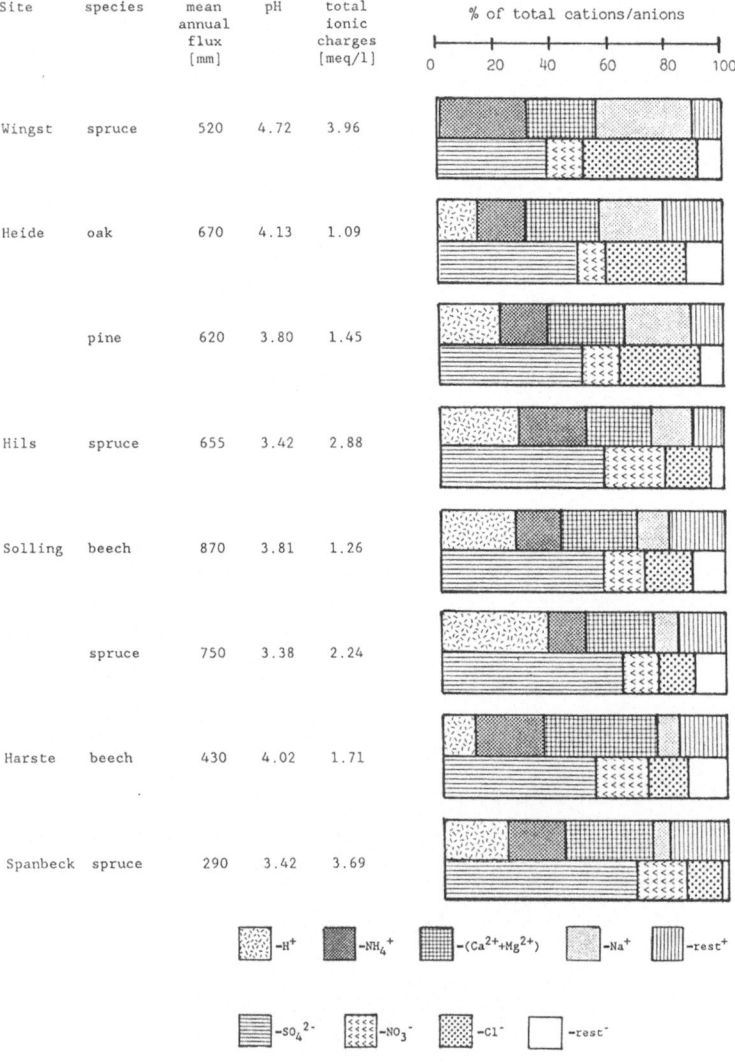

Fig. 3.   Solution chemistry of forest throughfall.

slightly higher value of 4.6 along with the highest total ionic charges in solution (calculated as cations + anions [meq^{-1}]). This is due to a pronounced sea salt influence in deposition, which becomes obvious from the solution composition at that site: Na$^+$ and Cl$^-$ are the greatest individual contributors to the cation and anion equivalent sums, in open precipitation as well as in throughfall.

A clear gradient in Na and Cl can be seen down the tabulation with increasing distance from the North Sea shore. This gradient is even more pronounced in throughfall than in open precipitation, indicating that interception of seaborne particles decreases rapidly with increasing distance from the shore.

While weighted mean Na : Cl-ratios are close to 1 : 1 on the equivalent base in open precipitation, they are consistently below 1 in throughfall at all sites. This can be either due to Cl-deposition from additional sources other than seawater (e.g., HCL from waste combustion, chemical industries) or to Cl$^-$-leaching from trees. There is, however, experimental evidence against substantial Cl$^-$-leaching from forest foliage (Fassbender, 1977; Hoefken, 1981).

The pH-range in throughfall is wider than in bulk rain. In seven of eight investigated forest ecosystems (Figure 3), long-term mean throughfall pH is lower than incident precipitation pH (ranging from 3.4 to 4.1). Three spruce stands, two high and one low elevation (Hils, Solling, Spanbeck), show very low throughfall pH (pH 3.4), whereas the deciduous sites in parallel case studies exhibit significantly higher values.

Only at the Wingst experimental site is the mean throughfall pH higher than precipitation pH. This site exhibits the highest sum of ionic charges and notably different solution composition, dominated by ammonium and Na on the cation side.

In all the forest stands, water fluxes to the ground are considerably lower than above canopy or in adjacent open plots. Total ionic charges in solution, on the other hand, are highly increased in throughfall. Interception and evaporation of water in the canopy increase the ion concentrations in throughfall. The relative increase in ionic content, however, is much larger than the relative decrease in water flux. Thus, a major fraction of the enhanced ion flux to the ground under forest stands is due to dry deposition or canopy sources. The increase in ionic strength on canopy passage is most pronounced at the spruce sites, as is the interception loss of water. This is a finding as would be expected, since both interception deposition of vapors, particles and cloud water and interception evaporation are functions of canopy structure and leaf surface area. It is generally understood that the latter is much higher for spruce stands compared to deciduous species (Ulrich et al., 1979). Moreover, deciduous canopies shed their leaves in the dormant season, thus effectively reducing surface area.

Looking at the patterns of ionic composition of open precipitation and throughfall (Figures 2 and 3), it can be stated that both fluxes are basically dilute solutions of H_2SO_4 and HNO_3 and their salts, plus a seasalt fraction, the influence of which is very much dependent on proximity to the coast. Sulfate is generally the major anion, except for the Wingst, as stated above. SO_4^{2-} fractions of total anionic charges range from ca. 35 to 50% in open precipitation, but usually account for more than 50% (up to 70%) in throughfall. This increased contribution of sulfate to the solution composition after

canopy passage (as well as the enhanced S-flux) indicates considerable interception deposition of S. Leaching of SO_4^{2-} from foliage is thought to be negligibly low, compared to
dry deposition rates of sulfur to the canopy (Ulrich *et al.*, 1979; Ulrich, 1983a; Fassbender,
1977; Hoefken, 1981).

In all case studies except Wingst, H^+ contributes a considerable fraction to the cation
sum in both rain and throughfall. Hydrogen ion in the very acid throughfall under Solling
spruce is almost 40% of the cation charge. Weighted mean throughfall at the Wingst
spruce site, on the other hand, exibits over an order of magnitude lower H^+ concentration which comprises only 1% of the total cation charge, while ammonium concentrations are very high. This indicates buffering of deposited acids by intercepted ammonia,
the neutralization product $((NH_4)_2 SO_4)$ appearing in throughfall or dry deposition of
primarily neutral ammonium sulfate salts (from NH_3/H_2SO_4 neutralization in the
atmosphere). High rates of NH_4 deposition are very likely in the Wingst area, since it
is located in a region of intensive livestock farming. Very high ammonia emissions and
deposition rates to forests have been found in similar situations in the Netherlands
(Roelofs *et al.*, 1985).

The Wingst data given in Figure 3 are an example of acid neutralization during canopy
passage, in this case by another deposited constituent (NH_3). Another example of net
acid removal from throughfall by a forest canopy can be seen in the Göttinger Wald case
study (Meiwes and König, 1986; Figure 4). The Göttinger Wald is a beech (*Fagus*

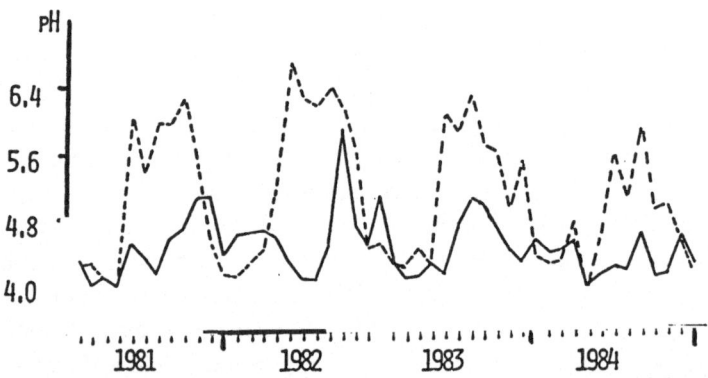

Fig. 4.   pH-values in incident precipitation (above canopy, solid line) and in throughfall (dashed line) in
a beech forest ecosystem on limestone in the Göttinger Wald, West Germany (from Meiwes and König,
1986).

sylvatica L.) forest ecosystem on limestone, which bears calcareous soils with high base
status. it can be seen from Figure 4, that under the condition of high soil alkalinity
buffering in the canopy can be very efficient and can even produce positive alkalinity
(pH > 5.5) in throughfall. The intensity of acid neutralization in the canopy follows a
typical seasonal pattern: during the leafless period buffering rates are zero or very low
and allow throughfall pH to fall below precipitation pH. When the canopy develops,

however, buffering begins and during the vegetation period throughfall occurs with an elevated pH.

The fact that at most of the sites shown in Figure 3 throughfall is more acidic than incident precipitation cannot be interpreted as evidence against proton buffering in the canopy. It only means that acid interception rates are higher than maximum buffering by canopy surfaces or dry deposited bases. Indications that H^+ buffering processes in the canopy are still active at sites with low throughfall pH are given by changes in ion ratios after canopy passage. At all of the sites in Figure 3, $(Ca^{2+} + Mg^{2+})$-concentrations in throughfall are considerably higher than in open precipitation, and $(Ca^{2+} + Mg^{2+})$-contributions to the cation equivalent sum have increased in most cases. The fraction of 'rest of positive charges' in throughfall is also higher. Most of this rest of cations is accounted for by K. Ca^{2+}, Mg^{2+}, and K^+ can all be involved in canopy internal H^+ buffering processes, the first two ions mostly in exchange reactions from inner surfaces, the latter during protonation of organic weak acid anions, that K^+ was bound to (Hofman et al., 1980).

At most of the forest sites the H^+/SO_4^{2-}-ratio in throughfall has decreased, compared to incident precipitation, while alkali earth cations from exchange processes and K^+ from weak acid titration occur in throughfall $((Ca^{2+} + Mg^{2+} + K^+)/SO_4^{2-}$-ratios increase. For instance, the calculated average for H^+/SO_4^{2-} in precipitation of all sites is 0.42, in throughfall it is 0.36. The respective values for Ca^{2+}/SO_4^{2-} are 0.28 for incident precipitation and 0.35 for throughfall).

Exceptions in this respect are the experimental sites Harste and Spanbeck, where Ca and Mg contents in open precipitation are extraordinarily high (Figure 2). While all other sites are situated in widespread forest areas, Harste and Spanbeck lie in an agricultural landscape around Göttingen. The high Ca^{2+}- and Mg^{2+}-fluxes in open land precipitation are due to soil dust sedimentation into the bulk samplers. This is similarly reflected in the $Ca^{2+} + Mg^{2+}$-levels in throughfall at these sites.

TABLE II

SO_4^{2-}–S fluxes to the ground in open precipitation and throughfall $(kg \cdot ha^{-1} \cdot a^{-1})$

	Wingst spruce	Heide		Harz spruce	Hils spruce	Solling		Harste beech	Spanbeck spruce
		oak	pine			beech	spruce		
PD	16.2	16.9	16.9	22.4	15.1	23.4	23.4	13.0	14.7
TF	61.6	28.6	36.0	42.7	86.6	50.3	85.2	31.3	57.3

In Tables II to IV mean annual fluxes of SO_4^{2-}–S, H^+, and Ca^{2+} in precipitation deposition (PD) and throughfall (TF) are listed (along with calculated total deposition and canopy interaction rates as discussed further below).

Hydrogen ion fluxes in open precipitation are highest in the mountaineous regions (Harz mountains, Hils, Solling). This is, however, mostly due to higher water flux at these elevated sites (Table I), since the acidity of incident rainfall shows no big regional differences (see Figure 2). H^+-fluxes in the open range from 0.2 (Wingst) to 0.8 kmol $\cdot$ ha$^{-1}\cdot$ a^{-1} (Harz, Solling).

Much higher H^+ fluxes are observed in throughfall at most of the sites. The maximum is held by Solling spruce with 3.1 kmol $\cdot$ $H^+\cdot$ ha$^{-1}\cdot$ a^{-1}. All coniferous sites except Wingst have H^+-fluxes in throughfall higher than 1 kmol $\cdot$ ha$^{-1}\cdot$ a^{-1}. The deciduous sites in lower elevations (Heide oak and Harste beech, see Table I) show considerably lower H^+ flux rates (0.5 and 0.4 kmol $\cdot$ ha$^{-1}\cdot$ a^{-1}, respectively), while Solling beech, a case study of a deciduous forest ecosystem in an exposed situation (Table I), shows moderately high H^+-flux in throughfall (1.3 kmol $\cdot$ ha$^{-1}\cdot$ a^{-1}).

There is an apparent relationship between the geographical situation and elevation of a forest ecosystem and its deposition rates of air pollutants. The mountain areas, especially Solling and Hils, which are frequently hit by air masses from the Rhein/Ruhr industrial region (Wilmers and Ellenberg, 1986), apparently receive the highest load of atmospheric acidity, as reflected in both H^+ and SO_4^{2-}-fluxes in forest throughfall. In addition to this regional differentiation of fluxes, however, an ecosystem-related differentiation also exists. The state of forest ecosystems can considerably influence

TABLE III

H^+ fluxes to the ground (open precipitation and throughfall) and calculated total deposition to the forest canopy (kmol $\cdot$ ha$^{-1}\cdot$ a^{-1})

	Wingst spruce	Heide		Harz spruce	Hils spruce	Solling		Harste beech	Spanbeck spruce
		oak	pine			beech	spruce		
PD	0.2	0.4	0.4	0.8	0.5	0.8	0.8	0.3	0.4
TF	0.1	0.5	1.0	1.3	2.5	1.3	3.1	0.4	1.1
TD	–	1.0	1.2	1.8	3.6	1.9	3.8	1.3	2.7
BF	–	0.5	0.2	0.5	1.1	0.6	0.7	0.9	1.6

PD = precip. dep., TF = throughfall, TD = calc. total dep., BF = buffering of H^+.

TABLE IV

Ca^{2+} fluxes to the ground (open precipitation and throughfall) and calculated total deposition and canopy leaching (CL) (kg $\cdot$ ha$^{-1}\cdot$ a^{-1})

	Wingst spruce	Heide		Harz spruce	Hils spruce	Solling		Harste beech	Spanbeck spruce
		oak	pine			beech	spruce		
PD	5.8	5.0	5.0	6.4	5.4	10.0	10.0	10.3	9.9
TF	26.7	12.8	17.4	19.8	33.6	24.1	32.5	23.3	27.2
TD	14.0	6.3	8.0	9.5	7.7	17.5	21.1	14.6	16.2
CL	12.7	6.2	9.4	10.3	25.9	6.6	11.4	8.7	11.0

atmospheric deposition rates. Main parameters of ecosystem state in this respect are species composition, age (particularly tree height), structure and density. Species related differentiations of solution acidity and flux rates have already been demonstrated in some examples above. The influence of stand age and orographic exposure can be seen from the Harz spruce site data. Although the higher Harz Mountains are a very exposed region (elevations 600 to 900 m), this spruce forest ecosystem exhibits moderately low fluxes of H^+ and SO_4^{2-}–S (Tables II and III). This can be explained by its location in a central Harz valley (elev. 700) and the younger stand age (Table I). The uniformity of a young, dense canopy decreases its aerodynamic roughness and wind eddy penetration.

A synopsis of the flux data and the site characteristics gives the impression that time averaged atmospheric deposition into forests might be controlled by a few determining regional and ecosystem characteristics.

4. Estimation of Total Deposition Rates and Canopy Interactions

The approach to estimate total deposition from the canopy flux balance described here has been gradually developed in forest ecosystem monitoring studies over the past 15 yr (Mayer and Ulrich, 1974; Ulrich *et al.*, 1979; Ulrich, 1983a). It is meant for the assessment of time averaged total deposition and canopy exchange rates and can be used with very simple experimental installations in the field. Only bulk collectors for open precipitation and throughfall (in a sufficient number of replicates to account for spatial variability) are needed (see methods above).

It must be clearly stated that experimental design and calculations are aimed at determining long-term trends, i.e., over years or longer periods. Inferences about air pollutant concentrations and their temporal variability, and about short-term deposition rates and peaks, cannot be made. In addition, the method rests on several key assumptions, as discussed below.

With the approach, a slightly different terminology from meteorological definitions is used. Figure 5 shows the partition of total deposition, as used for the calculations below.

Total deposition (TD) consists of precipitation deposition (PD) and interception deposition (ID). The point to distinguish these is not whether deposition occurs in wet or dry form, but if it occurs independently from the receptor surface by gravitational settling or if it is dependent on the filtering efficiency of the receptor (area, structure, chem. affinity, etc.).

The idea of the canopy flux balance is to use the natural forest surface as the receptor to sample atmospheric deposition. Net throughfall (i.e., the enrichment of element flux under a canopy, TF–PD) has then to be partitioned into interception deposition (ID, Figure 5) and a canopy sink/source component ($\pm$ S). This is expressed in the following formulas:

$$TD_x = PD_x + ID_x ; \quad x = \text{chem. element} \tag{1}$$

$$TF_x = PD_x + ID_x \pm S_x \tag{2}$$

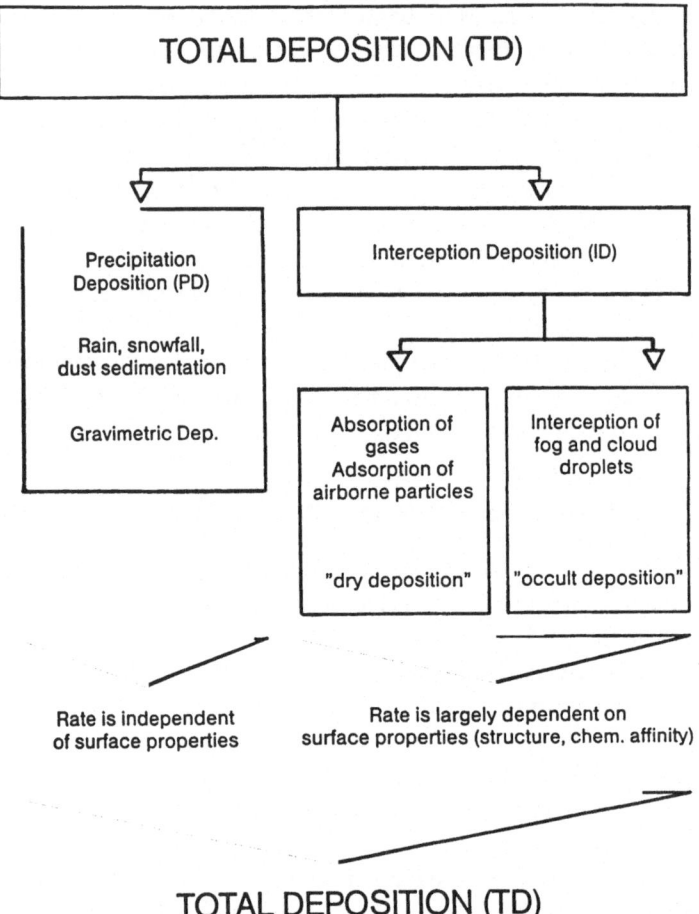

TOTAL DEPOSITION (TD)

Fig. 5.   Partition of total atmospheric deposition as used to calculate total deposition and canopy exchange rates from bulk solution samples (Equations (1) to (9)).

Equation (2) is valid if throughfall (TF) is the only relevant output flux from the canopy (i.e., gaseous or particulate releases from the leaves into the atmosphere are negligible, which can be assumed for the ionic constituents of interest here; TF in Equation (2) and the following equations is *total* throughfall, i.e. including stemflow). Stemflow can be substantial beech stands (up to 25 to 30% of total water flux), whereas it is almost negligeable in spruce stands (<5% of TF, Ulrich *et al.*, 1979).

Calculation of interception deposition (ID) starts with those elements, for which the canopy can be assumed to behave like an 'inert sampler', i.e., that have no substantial sinks/sources in the foliage layer. For these elements, ID is equal to net throughfall:

$$ID_x = TF_x - PD_x; \qquad x = Na, Cl, S.\eqno(3)$$

It is assumed that for Na, Cl, and S canopy exchange rates are negligibly low compared to atmospheric deposition inputs (Fassbender, 1977; Hoefken, 1981; Matzner, 1984,

1986; Lindberg *et al.*, 1986). Some studies, however, have lead to different conclusions for Na (Reiners and Olson, 1984) and for S (Lovett and Lindberg, 1984).

Assuming the validity of Equation (3), Na can be used as a 'tracer' to assess *particulate* ID of other elements. Particulate interception deposition, in this sense, includes cloud water or fog droplets.

$$\left(\frac{ID}{PD}\right)_{xpart} = \left(\frac{ID}{PD}\right)_{Na} ; \qquad xpart = H, K, Ca, Mg, Mn, S , \tag{4}$$

$$ID_{xpart} = \left(\frac{ID}{PD}\right)_{Na} PD_{xpart} . \tag{4a}$$

The assumption underlying Equation (4) is, that the interception ratio of Na (which is only deposited in particles) is equal to the *particulate* interception ratios of other elements. This assumption has, of course, no general experimental proof. It is meant as an approximation to estimate the particulate ID for other metals that are largely involved in canopy transformations (K^+, Ca^{2+}, Mg^{2+}, Mn^{2+}), whose atmospheric deposition rates can therefore not be inferred from throughfall.

For S, H, and N dry deposition in the gas phase (SO_2, HNO_3, NH_3) can play an important role. The total deposition of S has been assumed to be represented in S-flux in throughfall (Equation (3))*. Particulate ID can be estimated from Equation (4a). With this information, gaseous ID of S is:

$$ID_{Sgas} = TF_S - ID_{Spart} - PD_S . \tag{5}$$

Under the pH and redox conditions usually found in precipitation samples, most of the deposited SO_2 will be oxidized according to:

$$SO_2 + H_2O + Oxidant-O \rightarrow SO_4^{2-} + 2H^+ .$$

Thus, the calculated S_{gas} deposition can be assumed to represent an equivalent deposition of protons in most cases except in highly acidic solutions.

$$IS_{Sgas} = ID_{Hgas}^+ [eq] . \tag{6}$$

Total deposition of H^+ is therefore estimated as:

$$TD_H^+ = PD_H^+ + ID_{Hpart}^+ + ID_{Hgas}^+ . \tag{7}$$

As mentioned above, H^+-ions deposited to the canopy can be buffered by cation (esp. Ca^{2+}, Mg^{2+}) exchange from leaf tisue, or by weak organic acid titration**.

---

* This assumption entails the additional assumption that dry deposited SO_2 can be completely removed by subsequent rain events.
** This calculation assumes no significant deposition of HNO_3 vapor and that free H^+ in rain is not consumed by reactions with basic materials dry deposited to the bulk collector.

In these cases, deposited H^+ do not appear as measurable acidity in throughfall. The total H^+ load to the ecosystem, however, is not reduced by this buffering (see below, Figure 6).

1. Buffering of protons on inner leaf surfaces:

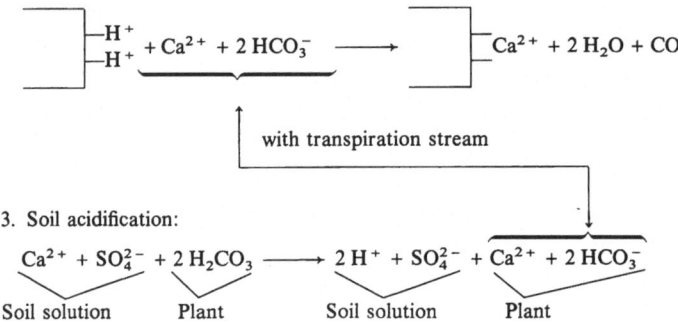

Leaching and
input to soil

2. Recharging the buffer:

with transpiration stream

3. Soil acidification:

$$Ca^{2+} + SO_4^{2-} + 2 H_2CO_3 \longrightarrow 2 H^+ + SO_4^{2-} + Ca^{2+} + 2 HCO_3^-$$

Soil solution   Plant   Soil solution   Plant

Fig. 6.   Schematic reaction pathway for the buffering of H^+ in a canopy and subsequent transfer of acidity to the rhizosphere (adapted from Ulrich, 1983b).

H^+-buffering rates in the canopy and leaching rates of cations can be estimated as follows:

$$H^+\text{-buffering} = TD_H^+ - TF_H^+ \tag{8}$$

$$CL_x = TF_x - TD_x ; \qquad x = Ca^{2+}, Mg^{2+}, K^+, Mn^{2+}$$
$$CL = \text{'canopy leaching'} \quad \text{(quantitatively important)} \tag{9}$$

The set of equations to calculate total deposition and canopy exchange of major elements has now been described. Some example results of using this approach are given in Tables II–IV, followed by a general discussion (chapter below).

For S TD rates are assumed to be approximately equal to flux in throughfall Equation (3). The site differentiation of this flux has already been discussed. Elevated, exposed sites with older spruce forests receive a maximum of atmospheric S-Deposition (Solling and Hils), while younger stands and deciduous forests in less exposed situations show considerably lower rates (Harz, Harste, Heide oak). Looking at the calculated H^+ total deposition values in Table III, the differentiation of sites comes out almost the same (this is, of course, partly due to interrelations in the calculation mode). Sulfate flux in throughfall is a good indicator of atmospheric acid deposition to a forest ecosystem. The H^+–TD rates calculated from the model, however, are generally higher than H^+

flux in throughfall. They range from 1 to 4 $kmol \cdot ha^{-1} \cdot a^{-1}$ in the long-term mean for the sites investigated. While H^+-flux in throughfall would underestimate total deposition to the canopy by a factor of 1.5 to 3, precipitation H^+ would be an underestimate of a factor 2 to 8.

Taking H^+ transport rates in open precipitation or throughfall as total deposition and estimating regional input on that basis must, therefore, lead to serious underestimates of acid deposition, as also reported elsewhere (Lindberg et al., 1986).

As an example of a cation involved in canopy exchange, calculated TD and leaching rates of Ca^{2+} are given in Table II. It can be seen that leaching from the canopy is an important component of Ca^{2+} flux to the ground. Calculated values are usually of the same magnitude as those for total deposition*.

5. Discussion

5.1. QUANTIFICATION APPROACH

The consideration that forests may interact with deposited constituents in terms of sink or source functions of the canopy is widely accepted in deposition research. To account for this, net throughfall has to be partitioned somehow into an interception deposition component (including dry particles, vapors, and cloud/fog water) and a canopy exchange component (including uptake or release).

If the existence of H^+ buffering mechanisms in foliage is accepted (Ulrich, 1983b), total H^+-deposition must amount to somewhere between the measured H^+ flux in throughfall at one extreme and the equivalent of strong acid anions ($SO_4^{2-} + NO_3^-$) at the other. This of course assumes that all deposited anions appear in throughfall, which is not as good an assumption for NO_3^- as it is for SO_4^{2-} (Lindberg et al., 1986). A maximum estimate of H^+ deposition would be the equivalent of non-marine sulfate plus nitrate in throughfall. However, other neutral sulfates and nitrates than seaborne particles can be deposited, and nitrate might undergo (like all other deposited N-compounds) considerable canopy transformations (Matzner, 1986), suggesting the total flux of H^+ generally to be less than the total flux of $SO_4^{2-} + NO_3^-$-equivalents in throughfall. Thus, a method to estimate particulate H^+ (mostly H_2SO_4, HNO_3-droplets in fog and cloud water) and particulate SO_4^{2-}-deposition independently plus a gaseous S deposition component seems appropriate. Results can be tested easily for fitting between the upper and lower limit of total H^+ as described above.

The assessment approach discussed here, however, rests on some crucial assumptions that can not be experimentally verified with the data currently available from our studies. Moreover, the experimental design may be subject to some uncertainties in the determination of fluxes due to reliance on bulk samplers which have a relatively long exposure in the field. But even if the approach can not be regarded as

---

* It must be kept in mind in these analyses that underestimates of dry deposition naturally lead to overestimates of canopy leaching. Hence if the true ratio of (ID/PD) for Ca is not the same as this ratio for Na because of source strength, particle size, or chemical (e.g., hygroscopicity) effects, the values of both PD and CL for Ca^{2+} need to be adjusted.

an absolutely accurate quantitative determination of all total deposition components, it can still be useful in the assessment of atmospheric element input and canopy transfer and their relative spatial patterns.

Experimental studies support the assumption that Na^+ is not leached from trees (Fassbender, 1977; Höfken, 1981). This is in good agreement with plant physiological considerations, since Na is not a major plant nutrient and should therefore not be taken up selectively and cycled by a forest. A more crucial assumption is that the wet and dry deposition characteristics of Na-particles (including Na-bearing droplets) represent all other particulate matter and cloud droplet interception. This assumption can only be regarded as an approximation. Another assumption, that might be crucial, is that S is not leached from forest canopies. The above mentioned experimental leaching studies (Fassbender, 1977; Hoefken, 1981), however, support it as well as recent field studies with application of ^{35}S to the xylem stream (Garten, 1986).

For the purposes of the calculations pursued here, it would be sufficient if SO_4^{2-} leaching did not exceed a minor fraction of total S in throughfall. This can be assumed with good confidence under central European conditions, where S emissions and deposition rates are high (Semb, 1978; Nodop, 1987) and where internal plant pools of SO_4^{2-} are comparatively low (Meiwes, 1979). Ulrich (1983a) estimates a mean value of ca. 1.5 (0–5) $kg \cdot ha^{-1} \cdot a^{-1}$ for S leaching from senescent leaves of a beech stand in northwestern Germany (Solling). The corresponding mean S flux in throughfall is 52 $kg \cdot ha^{-1} \cdot a^{-1}$. This leaching fraction (ca. 5% of total flux) would be a tolerable error in the calculations.

5.2. TOTAL DEPOSITION AND CANOPY EXCHANGE

As stated above, the differentiation of sites in the case studies reported here comes out the same either by looking at (observed) $SO_4^{2-} - S$-fluxes in throughfall or by calculating total H^+ deposition in the way described (Equations (1) to (9); Table III). It seems reasonable that total sulfate in throughfall should be related to total deposited acidity in industrialized regions.

Acidity deposition rates seem to be related to a few determining factors, which could be generally described as regional characteristics (proximity and exposure to sources) and ecosystem characteristics (species, age, stand density, canopy structure). In this respect, the assessment model yields plausible results.

The data in Tables III and IV show that canopy interactions play an important role in determining throughfall fluxes with respect to some major ionic constituents. The most important canopy transformations of deposited constituents are (besides of direct assimilation of nutrients, e.g. N) H^+-buffering in the canopy and metal cation leaching.

These canopy effects must be considered when estimating total acidity deposition or deposition rates of metals. The method of assessment proposed here is thought to help in this case, by means of relatively simple field installations.

An important fact to keep in mind is that H^+ buffering in the canopy removes acidity from throughfall, but does *not* decrease the total H^+ load to soil. Ulrich (1983b) has described the mechanism of H^+-transfer to the rhizosphere following H^+-buffering in the canopy (Figure 6). This mechanism assumes buffering to occur by cation exchange from inner leaf tissue surfaces. Buffering by weak acid anions with subsequent leaching to the forest floor is not considered in these reactions, but would have the same effect when recharging buffer capacity.

To regenerate the buffer systems, equivalent amounts of alkali and alkali earth cations to the buffer capacity consumed in the canopy have to be taken up from soil solution. To maintain charge balance, the amount of H^+ previously buffered in the crowns is now released to the soil at the root surface. Altogether, H^+-buffering in the canopy is just a re-routing of acidity from throughfall into soil solution and soil solid phases.

Calibration of regional H^+ deposition models has to regard this fact. An assessment of *total* H^+ input to terrestrial ecosystems is needed, open precipitation or throughfall H^+-fluxes will usually be considerable underestimates of H^+ ion load to soil.

Acknowledgments

The author wishes to thank Dr Steven E. Lindberg, visiting scientist from Oak Ridge National Laboratories, for his careful review and constructive, critical discussions on this paper.

References

Boynton, D.: 1954, *Ann. Rev. Plant. Physiol.* **5**, 31.

Bredemeier, M.: 1987, 'Stoffbilanzen, interne Protonenproduktion unc d Gesamtsäurebelastung des Bodens in verschiedenen Waldökosystemen Norddeutschlands', Dissertation Univ. Göttingen.

Büttner, G., Lamersdorf, N., Schultz, R., and Ulrich, B.: 1986, *Deposition und Verteilung chemischer Elemente in küstennahen Waldstandorten*, Ber. d. Fz. Waldökosys./Waldsterben, Univ. Göttingen, Reihe B Bd. 1.

Fassbender, H. W.: 1977, *Ecologia plantarum* **12**, 263.

Garten, C. T.: 1986, 'Sulfur Isotope Studies on Whole Trees', in S. I. Auerbach (ed.), *Annual Report of the Environmental Sciences Division 1986*, ORNL, Oak Ridge, U.S.A.

Godt, J.: 1986, *Untersuchung von Prozessen im Kronenraum van Waldökosystemen*, Ber. d. Fz. Waldökosys./Waldsterben, Univ. Göttingen Bd. 19, pp. 1–265.

Hauhs, M.: 1985, *Wasser- und Stoffhaushalt im Einzugsgebiet der Langen Bramke (Harz)*, Ber. d. Fz. Waldökosys./Waldsterben, Univ. Göttingen Bd. 17, pp. 1–206.

Höfken, K. D.: 1981, 'Untersuchungen über die Deposition atmosphärischer Spurenstoffe an Buchen- und Fichtenwald', Dissertation, Institut für Meteorologie und Geophysik der J. W. Goethe-Univ. Frankfurt.

Hofman, W. A., Lindberg, S. E., and Turner, R. R.: 1980, *J. Env. Qual.* **9**, 95.

Lindberg, S. E., Lovett, G. M., Richter, D. D., and Johnson, D. W.: 1986, *Science* **231**, 141.

Lovett, G. M. and Lindberg, S. E.: 1984, *J. Appl. Ecol.* **21**, 1013.

Lutz, H. J.: 1987, 'Einfluß von saurem Nebel auf die Ausbildung von Schadsymptomen bei jungen Fichten', Dissertation Univ. Gießen, F.R.G., 128 p.

Matzner, E.: 1984, *Deposition und Umsatz chemischer Elemente im Kronenraum von Waldbeständen*, Ber. d. Fz. Waldökosys./Waldsterben Univ. Göttingen Bd. 2, pp. 61–87.

Matzner, E., Khanna, P. K., Meiwes, K. J., Gassens-Sasse, E., Bredemeier, M., und Ulrich, B.: 1984, *Ergebnisse der Flüssemessungen in verschiedenen Waldökosystemen*, Ber. d. Fz. Waldökosys./Waldsterben Univ. Göttingen Bd. 2, pp. 29–49.

Matzner, E.: 1986, 'Deposition/Canopy – Interactions in Two Forest Ecosystems of Northwest Germany', in H. W. Georgii (ed.), *Atmospheric Pollutants in Forest Areas*, D. Reidel Publ. Co., Dordrecht, Holland, pp. 247–262.

Mayer, R. and Ulrich, B.: 1974, *Oecol. Plant.* **9**, 157.

Meiwes, K. J.: 1979, *Gött. Bdkdl. Ber.* **60**, 1.

Meiwes, K. J. und König, N.: 1986, *H-Ionen-Deposition in Waldökosystemen in Norddeutschland*, GSF München, BPT-Bericht 8/86, pp. 25–35.

Nodop, K.: 1987, 'Nitrate and Sulfate Wet Deposition in Europe', in G. Angeletti and G. Restelli (eds.), *Physico-Chemical Behaviour of Atmospheric Pollutants*, D. Reidel Publ. Co., Dordrecht, Holland, pp. 520–528.

Reiners, W. A. and Olson, R. K.: 1984, *Oecologia* **63**, 320.

Roelofs, J. G. M., Kempers, A. J., Hondijk, A. L. F. M., and Jansen, J.: 1985, *Plant and Soil* **84**, 45.

Semb, A.: 1978, *Atm. Env.* **12**, 455.

Streletzki, H. W.: 1986, *Forst- und Holzwirt* **41**, 541.

Ulrich, B., Mayer, R., und Khanna, P. K.: 1979, *Die Deposition von Luftverunreinigungen und ihre Auswirkungen in Waldökosystemen im Solling*, Schriften aus der Forstl. Fak. d. Univ. Göttingen, Bd. 58, Sauerländer-Verlag.

Ulrich, B.: 1983a, 'Interaction of Forest Canopies with Atmospheric Constituents: SO_2, Alkali and Earth Alkali Cations and Chloride', in B. Ulrich and J. Pankrath (eds.), *Effects of Accumulation of Air Pollutants in Forest Ecosystems*, D. Reidel Publ. Co., Dordrecht, Holland, pp. 33–45.

Ulrich, B.: 1983b, 'A Concept of Forest Ecosystem Stability and of Acid Deposition as Driving Force for Destabilization', in B. Ulrich and J. Pankrath (eds.), *Effects of Accumulation of Air Pollutants in Forest Ecosystems*, D. Reidel Publ. Co., Dordrecht, Holland, pp. 1–29.

Ulrich, B. (ed.): 1986, *Raten der Deposition, Akkumulation und des Austrags toxischer Luftverunreinigungen als Maß der Belastung und Belastbarkeit von Waldökosystemen*, Ber. d. Fz. Waldökosys./Waldsterben, Univ. Göttingen, Reihe B, Bd. 2, pp. 1–210.

Wilmers, F. und Ellenberg, H.: 1986, in H. Ellenberg, R. Mayer, and J. Schauermann (eds.), *Ökosystemforschung – Ergebnisse des Solling-Projekts*, Ulmer, Stuttgart, pp. 61–76.

List of Participants

Joseph Alcamo,
International Institute for
   Applied Systems Analysis,
Schlossplatz 1,
A-2361 Laxenburg,
Austria.

Jerzy Bartnicki,
International Institute for
   Applied Systems Analysis,
Schlossplatz 1,
A-2361 Laxenburg,
Austria.

Michael Bredemeier,
Forschungszentrum Waldökosysteme
   der Universität Göttingen,
D-3400 Göttingen, F.R.G.

Gregory R. Carmichael,
University of Iowa,
129 Chem Bldg.,
Iowa City, IA 52240,
U.S.A.

Charles Hakkarinen,
Electric Power Research Institute
Environmental Data Analysis Dept.,
P.O. Box 10412,
Palo Alto, CA 94303,
U.S.A.

Leen Hordijk,
National Institute for Public
   Health & Environmental Hygiene,
Laboratory for Waste Research
   and Emissions,
P.O. Box 1,
3720 BA Bilthoven,
The Netherlands.

Jan Juda
Institute of Environmental Eng.,
Warsaw Technical University,
ul. Nowowiejska 20,
00–653 Warsaw,
Poland.

Katarzyna Juda,
Institute of Environmental Eng.,
Warsaw Technical University,
ul. Nowowiejska 20,
00–653 Warsaw,
Poland.

Juha Kämäri,
National Board of Waters,
Water Research Institute,
Pohjoinen Rautatienkatu 21B,
SF-00100 Helsinki,
Finland.

Werner Klug,
Technische Hochschule,
University of Darmstadt,
Institute for Meteorology,
Hochschulerstrasse 1,
D-6100 Darmstadt,
F.R.G.

Ralph Lehmann,
Meteorological Service of the GDR,
Albert-Einstein Strasse 42–44–46,
Potsdam 1561,
G.D.R.

Annikki Mäkelä,
Department of Silviculture,
University of Helsinki,
Unioninkatu 40B,
00170 Helsinki,
Finland.

Robert E. Munn,
Institute for Applied Systems
  Analysis,
Schlossplatz 1,
A-2361 Laxenburg,
Austria.

Brand L. Niemann,
Office of Air and Radiation,
U.S. Environmental Protection
  Agency,
Washington, DC 20460,
U.S.A.

Goran Nordlund,
Finnish Meteorological Institute,
Vuorikatu 24,
P.O. Box 503,
10–00101 Helsinki,
Finland.

Sergei Pitovranov,
Institute for Applied Systems
  Analysis,
Schlossplatz 1,
A-2361 Laxenburg,
Austria.

Jerzy Pruchnicki,
World Meteorological Organization,
41, Av. Giuseppe-Motta,
1201 Geneva 20,
Switzerland.

Jørgen Saltbones,
Norwegian Meteorological Institute,
P.O. Box 320,
Blindern, Oslo 3,
Norway.

Deszo J. Szepesi,
Institute for Atmospheric Physics,
P.O. Box 39,
H-1675 Budapest,
Hungary.

Joop F. den Tonkelaar,
Eykmannweg 1,
3731 KT de Bilt,
The Netherlands.

Marek Uliasz,
Institute of Environmental Eng.,
Warsaw Technical University,
ul. Nowowiejska 20,
00–653 Warsaw,
Poland.

List of Presentations

J. Alcamo. Results from an uncertainty analysis of the EMEP I model of sulfur transport in Europe.

J. Bartnicki. Preliminary analysis of parameter uncertainty of the EMEP II model of sulfur transport in Europe.

M. Bredemeier. Forest canopy transformation of atmospheric deposition.*

S.-Y. Cho, G. Carmichael, H. Rabitz. Relationships between primary emissions and acid deposition in Eulerian models determined by sensitivity analysis.*

U. Damrath, R. Lehmann. On coupling air pollution transport models of different scales.*

W. Klug, B. Erbschäusser. Application of the FAST method to a long-term interregional air pollution model.*

B. L. Niemann. Forecasted sulfur depositions in Europe: results from a personal computer-regional climatological deposition model and data analysis.*

G. Nordlund. Aspects of using long range transport information in a regional scale model of sulfur and nitrogen oxides in Finland.

S. E. Pitovranov. The assessment of impacts of possible climatic change on the results of the EMEP and IIASA sulfur deposition models of Europe.*

D. Szepesi. Considerations of including long range transmission of air pollutants in a regional model of Hungary.

J. den Tonkelaar. The effect of future climatic change on sulfur deposition in Europe.

M. Uliasz. Application of the FAST method to sensitivity-uncertainty analysis of a Lagrangian model of sulfur transport in Europe.*

* Written paper included in this volume of *Water, Air, and Soil Pollution.*